Anatomy and Physiology for Students

A College level Study Guide for Life Science and Allied Health Majors

Leonel Travers

Chapter 1: Review of high school sciences that relate to anatomy and physiology.

Courses on anatomy and physiology usually build on concepts taught in high school chemistry and biology classes. The purpose of this chapter is to review and highlight the foundational concepts that will help you ace your anatomy and physiology course. In addition to this we will also cover the language of anatomy.

Terminologies

Anatomy is a branch of science that deals with the structure of living organisms.

Adenosine triphosphate also called ATP is an organic molecule that stores energy in the cell. In addition to being an energy source for biological reactions ATP also acts as a hydrotrope, which simply means it "turns towards water". This property of ATP allows it enhance protein solubility.

Atom is the smallest unit of a chemical element.

A **bond** or **chemical bond** is the lasting interaction between different atoms, molecules or ions. **Ionic** and **covalent bonds** are examples of strong bonds while **hydrogen bonds** and **van der Waals interactions** are examples of weak bonds. An **ionic bond** is the electrostatic force of attraction between two oppositely charged ions. Table salt or sodium chloride has an ionic bond between a sodium ion and a chloride ion. **Covalent bond** is binding that arises due to sharing an electron pair between two atoms. Elements of biological molecules in our cells usually form covalent bonds. **Hydrogen bonds** are a type of interaction between an electronegative atom and a hydrogen atom bonded to another electronegative atom. Hydrogen bonds are responsible for holding together two strands of DNA. van der Waals interaction is caused by temporary and spontaneous attractions between electron-rich regions of one molecule and electron-poor regions of another. **Nonpolar bonds** are formed between two atoms that share their electrons equally while **polar bonds** form when two bonded atoms share electrons unequally.

Carbohydrates are one of three main classes of energy source in our diet. They are organic compounds and can be considered as hydrates of carbon. The ratio of carbon to hydrogen to oxygen in carbohydrates is 1:2:1. Carbohydrates can be classified nutritionally or based on their chemistry.

Nutritionally, three main types of carbohydrates are **sugars**, **starches** and **fiber**. Sugars are also called simple carbohydrates. Starches are complex carbohydrates, which are made of lots of simple sugars strung together. Your body can break down starches into simple sugars and use them for energy. Fibers are also complex carbohydrates but unlike starches our body cannot breakdown most fibers. So fibers give us a sense of satiety, the satisfied feeling of being full after eating.

Based on their chemical structure, carbohydrates can be classified as **monosaccharides**, **disaccharides** and **polysaccharides**. Monosaccharide is a simple sugar consisting only of one sugar unit. Glucose or fructose are a couple of examples of simple sugar. Disaccharides are a group of sugars composed of two monosaccharide groups linked together. The link between two monosaccharides to form a disaccharide is called a glycosidic linkage. Table sugar also called sucrose is a disaccharide and is made from one molecule of glucose and fructose each. Polysaccharides are polymers of monosaccharides. They are the most abundant carbohydrate found in food. Starches and fibers are examples of polysaccharides.

Elements are substances made up of entirely one kind of atom. So they cannot be decomposed into other substances by chemical processes.

Homeostasis is the property of living systems to remain in a state of equilibrium and stability necessary for survival. We will learn more about this in chapter 2.

Lipids are any class of organic compounds that are fatty acids or their derivatives and are insoluble in water but soluble in organic solvents. This is because they are **hydrocarbons** that include mostly **nonpolar** carbon–carbon or carbon–hydrogen bonds. Major classes of lipids include **fats, oils, waxes, phospholipids**, and **steroids.**

Glycerol and **fatty acids** are the two main components of a fat or oil molecule. **Fatty acids** are attached to each of the **three** carbons of the **glycerol** molecule by an **ester bond** through an oxygen atom. This is why fats are also called **triglycerides.** Fatty acids are called **saturated** when there are only single bonds between neighboring carbons in the hydrocarbon chain, this means that the fatty acids are saturated with hydrogen. When at least a pair of carbons in the hydrocarbon chain have double or triple bonds then the fatty acid is called **unsaturated**. Fatty acids with triple bonds tend to be very bioactive are not well understood to have a metabolic role. So for the purposes of the rest of the chapter when we discuss unsaturated fatty acids we mean fatty acids with double bonds. If the fat has more than one double bond it is called **poly-unsaturated fatty acid**. Most unsaturated fatty acids are liquids at room temperature and are called oils. Unsaturated fats often contain **cis** unsaturated fatty acids. **Cis** and **trans** refer to the configuration of the molecule around the double bond. If hydrogens are present in the same plane, it is referred to as a **cis fat**; if the hydrogen atoms are on two different planes, it is referred to as a **trans fat**. **Essential fatty acids** are required for biological processes but cannot be synthesized in the human body, therefore they need to be supplemented through ingestion via the diet. **Omega-3** and **Omega-6 fatty acids** are the only two known essential fatty acids for humans.

Waxes are a type of long chain nonpolar lipid. Natural waxes are typically esters of fatty acids and long chain alcohols. They prevent water from sticking on the surface and for this reason are secreted on the feathers of some aquatic birds and the leaf surfaces of some plants.

Phospholipids are a type of lipid molecule that are the major component of cell membranes. Unlike triglycerides, which have three fatty acids, phospholipids have two fatty acids that help form a **diacylglycerol**. This forms the **non-polar tail** of a phospholipid. The third carbon of the glycerol backbone is occupied by a modified phosphate group. This is the **negatively-charged polar head**, which is hydrophilic. Since the heads are hydrophilic, they face outward and are attracted to the intracellular and extracellular fluid. If phospholipids are placed in water, they arrange themselves in a spherical form in aqueous solutions, this is called a **micelle**. When the phospholipids are arranged tail to tail in two adjacent sheets it is called a **phospholipid bilayer**. Cell membranes have a phospholipid bilayer with the hydrophobic tails associating with one another, forming the interior of the membrane. Because of the hydrophobic tails and the hydrophilic heads the lipid bilayer acts as a **semipermeable membrane** and separates the cell from its environment.

Steroids have a characteristic fused ring structure containing four rings of carbon atoms. They include many hormones, alkaloids, and vitamins.

Cholesterol is the most common steroid and is a precursor to many steroid hormones such as testosterone and estradiol. It is also a precursor for vitamin D.

Nucleic acids are biopolymers and their monomers are called **nucleotides** which are made up of three components: a nitrogenous base, a pentose sugar, and a phosphate group. The two main types of nucleic acids are **deoxyribonucleic acid (DNA)** and **ribonucleic acid (RNA)**. Each nucleotide in DNA contains one of four possible nitrogenous bases: **adenine** , **guanine**, **cytosine**, and **thymine**; while each nucleotide in RNA contains the same four possible nitrogenous bases as DNA except for **uracil** replacing **thymine**. Adenine and guanine are classified as **purines** with a primary structure consisting of two carbon-nitrogen rings. Cytosine, thymine, and uracil are classified as **pyrimidines** which have a single carbon-nitrogen ring as their primary structure.

DNA is the molecule inside cells that contains the genetic information responsible for the development and function of an organism. DNA molecules allow this information to be passed from one generation to the next. The central dogma of molecular biology states that the genetic information flows from DNA to RNA to proteins. We would like to introduce briefly the various kinds of RNA.

Messenger RNA or **mRNA** carries the genetic code from DNA in the nucleus to the cytoplasm in a form that can be recognized to make proteins. This is also called coding RNA since it codes for information.

The other kinds of RNA are the non-coding RNA

Ribosomal RNA or **rRNA** is the catalytic component of the **ribosomes**. In the cytoplasm, rRNAs and protein components combine to form a nucleoprotein complex called the ribosome which binds **mRNA** and synthesizes proteins.

Small nuclear RNAs or **snRNAs** are associated with specific proteins in the nucleus to form complexes called **small nuclear ribonucleoproteins** or **snRNP**. The primary function of **snRNPs** is to process the precursor **mRNA**.

Small nucleolar RNAs or **snoRNAs** are associated with specific proteins in the nucleolus to form complexes called **small nucleolar ribonucleoproteins** or **snoRNP**. The primary function of **snoRNPs** is in the maturation of **rRNA**.

Piwi-interacting RNA or **piRNA** is the largest class of small non-coding RNA molecules in animal cells. They bind the PIWI subfamily proteins that are involved in keeping the genome stable in germline cells.

Long noncoding RNAs or **lncRNA** are a heterogeneous group of non-coding RNA that modulates chromatin structure, function, transcription of genes, and affect RNA splicing, stability and translation.

MicroRNAs or **miRNA** control gene expression by binding with messenger RNA in the cell cytoplasm. The mRNA thus marked is destroyed and its components recycled.

Proteins are large macromolecules comprising of one or more long chains of amino acid residues. They can broadly serve five class of roles in the body - **structural**, **storage**, **hormonal**, **enzyme** and **immunoglobulins**. Structural proteins like keratin and collagen are essential for constructing the body. Storage proteins are used as a reservoir for critical elements, example hemoglobin stores oxygen. We will discuss more about these in chapter 5 on the integumentary system. Hormonal proteins act as chemical messengers. We will discuss more about these in chapter 6 on the endocrine system. Enzymes are biological catalysts of biochemical reactions. Immunoglobulins are antibodies that protect the body from pathogens.

Parts of a cell

Cytoplasm is a thick solution that fills each cell and is enclosed by the **cell membrane**. It is mainly composed of water, salts, and proteins.

Cell membrane is a thin semipermeable membrane that surrounds every living cell. It is mainly composed of phospholipids, steroids and proteins.

Nucleus is a membrane-bound organelle that contains the cell's chromosomes.

Nucleolus is a small, typically spherical granular body located in the nucleus of a eukaryotic cell, composed largely of protein and RNA.

Chromosomes are an organized package of DNA found in the nucleus of the cell. Humans have 23 pairs of chromosomes, 22 pairs of numbered chromosomes, called **autosomes**, and one pair of sex chromosomes, X and Y.

Organelles are any of the various cellular structures that perform a distinctive function inside a cell.

Ribosomes are the sites of RNA and associated proteins where information carried in the genetic code is converted into protein molecules.

Endoplasmic reticulum is a network of membranes inside a cell through which proteins and other molecules move. There are two kinds of endoplasmic reticulum – smooth and rough. The **rough endoplasmic reticulum** has ribosomes assembling cell membrane proteins attached on its membrane. **Smooth endoplasmic reticulum** lacks ribosomes and helps synthesize and concentrate lipids, phospholipids, and steroids.

Golgi bodies or the **Golgi apparatus** are an organelle in the cells with a major function of modifying, sorting and packaging proteins meant for secretion. It is also transports lipids around the cell, and is involved in creating the **lysosomes**. The sacs or folds of the Golgi body are called **cisternae**.

Lysosomes are acidic membrane-bound organelles containing numerous hydrolytic enzymes which catalyze hydrolysis reactions.

Mitochondria are double membrane-bound organelles in which the biochemical processes of respiration and energy production occur. It is also a site of calcium storage and heat generation.

The language of anatomy

Superior or **cranial** is a directional term that describes positions toward the head end of the body.

Inferior or **caudal** describes positions away from the head.

Anterior or **ventral** is a directional term that describes positions in the front of the body.

Posterior or **dorsal** describes positions in the back of the body.

Medial is a directional term that describes positions toward the midline of the body.

Lateral describes positions away from the midline of the body.

Proximal is a directional term that describes positions closest to the point of origin of a part.

Distal describes positions farthest away from the point or origin of a part.

Coronal plane or **frontal plane** describes a vertical plane running from side to side.

Sagittal plane or **lateral plane** is a vertical plane running from front to back.

Axial plane or **transverse plane** describes a horizontal plane dividing the body into upper and lower parts.

Median plane or **sagittal plane** is a plane through the midline of the body dividing it into right and left halves.

Thoracic cavity or **chest cavity** is a space in the body containing the heart, lungs, trachea, esophagus, large blood vessels, and nerves. It is protected by the thoracic wall consisting of the rib cage and associated skin and muscle.

Abdominal cavity is a space in the body that contains most of the gastrointestinal tract as well as the kidneys and adrenal glands.

Pelvic cavity is a space that contains the urogenital system as well as the rectum.

Dorsal cavity is a fluid filled space that surrounds the brain and spinal cord.

Human Body Systems

Though there is no standard or universally accepted number one can consider that the human body has approximately **79 organs**. These organs don't always function independently, instead they function in groups called **organ systems**.

An **organ system** consists of groups of organs working together to accomplish similar tasks. There are **11 organ systems** in the human body, these include the following.

Circulatory system which deals with the blood supply. We will cover this in detail in Chapters 8 and 11.

Respiratory system which deals with breathing works with the circulatory system to provide oxygen and to remove the waste products of metabolism. We will cover this in Chapter 9.

Digestive system which deals with digestion and absorption of nutrients from food. We will cover this in detail in Chapter 7.

Nervous system is one of two organ systems in the body that maintains homeostasis. We will cover homeostasis in Chapter 2 and nervous system in Chapter 12.

Endocrine system is the other organ system that regulates homeostasis. This organ system is a collection of glands that release hormones into the bloodstream. We will cover endocrine system in Chapter 10.

Excretory system deals with waste elimination. It is composed of the urinary, respiratory, integumentary, and digestive systems. We will cover urinary system in Chapter 10.

The **integumentary system** consists of the skin and the associated structures like glands and hair. We will cover it in Chapter 5.

Lymphatic system functions to drain tissue fluid, plasma proteins and other cellular debris back into the blood stream. It is also includes components of the immune system which are responsible for protecting the body. We will cover the lymphatic system in Chapter 11.

The **skeletal system** has six different functions, the most visible role of the system is to provide support to the body. Chapter 3 will provide more detailed information on the skeletal system.

The **muscular system** is involved in generating motion We will cover this in detail in Chapter 4. Some anatomy texts might combine the skeletal and muscle systems together into the **musculoskeletal system.**

Human Body Cavities

The cavities of the body have several functions, including housing the internal organs, shock absorption, compartmentalizing organs and limiting the spread of infections. There are **16 body cavities.**

Ventral body cavity is at the anterior, or front, of the trunk. Organs contained within this body cavity include the lungs, heart, stomach, intestines, and reproductive organs. The organs within the ventral cavity are also called the viscera.

Dorsal body cavity is a continuous cavity located on the dorsal side or back of the body. It contains the organs of the upper central nervous system, including the brain and the spinal cord.

Cranial cavity is the anterior or top portion of the dorsal cavity consisting of the space inside the skull. This cavity contains the brain, the meninges of the brain, and cerebrospinal fluid.

Vertebral canal is the most narrow of all body cavities and is the posterior or bottom portion of the dorsal cavity. It contains the vertebral column which includes the spinal cord, the meninges of the spinal cord, and the fluid-filled spaces between them.

Thoracic cavity is the anterior or top portion of the ventral body cavity and is found within the rib cage in the torso with the diaphragm at the bottom. It contains the primary organs of the cardiovascular and respiratory systems, such as the heart and lungs, and also other organs like the esophagus and the thymus gland.

Abdominopelvic cavity is the posterior or bottom portion of the ventral body cavity located below the thoracic cavity and diaphragm. It contains the major organs of the digestive, reproductive, and urinary systems.

Abdominal cavity is the anterior or top portion of the abdominopelvic cavity. It is not contained within bone and houses organs of the digestive and renal systems, as well as some organs of the endocrine system like the adrenal glands.

Pelvic cavity is the posterior or bottom portion of the abdominopelvic cavity. It is contained within the pelvis and houses the bladder and reproductive system.

The **mediastinum** is the compartment that runs the length of the thoracic cavity between the pleural sacs of the lungs. Although there are no physical barriers between compartments other than the pericardium, the mediastinum is typically subdivided into four compartments.

Superior mediastinum is essentially a conduit space allowing structures to pass between the head, neck, and thorax. It has the following structures

1. *Organs of the superior mediastinum:* thymus, trachea, and esophagus
2. *Arteries of the superior mediastinum:* aortic arch, brachiocephalic trunk, left common carotid artery, and left subclavian artery
3. *Veins and lymphatic vessels of the superior mediastinum:* superior vena cava, brachiocephalic veins, the arch of the azygos, and thoracic duct
4. *Nerves of the superior mediastinum:* left and right vagus, recurrent laryngeal, cardiac, left and right phrenic nerves

Inferior mediastinum is the space that houses the remaining three compartments of the mediastinum, namely, the **anterior, middle** and **posterior mediastinum.**

Anterior mediastinum is filled with connective and fatty tissue that cushions and supports the thymus as well as the vital cardiac structures just posterior to it. It has the following structures

1. *Organs of the anterior mediastinum:* thymus

2. ***Arteries of the anterior mediastinum:*** internal thoracic branches
3. ***Veins and lymphatic vessels of the anterior mediastinum:*** internal thoracic branches, and parasternal lymph nodes

Middle mediastinum is bounded by the pericardial sac and has the following structures

1. ***Organs of the middle mediastinum:*** the heart and its great vessel roots, trachea and main bronchi
2. ***Arteries of the middle mediastinum:*** ascending aorta, pulmonary trunk, pericardiacophrenic arteries
3. ***Veins and lymphatic vessels of the middle mediastinum:*** superior vena cava, pulmonary veins, pericardiacophrenic veins
4. ***Nerves of the middle mediastinum:*** phrenic, vagus, and sympathetic nerves

Posterior mediastinum provides space for the passage between the thoracic and abdominal cavities. It has the following structures

1. ***Organs of the posterior mediastinum:*** esophagus
2. ***Arteries of the posterior mediastinum:*** descending thoracic aorta
3. ***Veins and lymphatic vessels of the posterior mediastinum:*** azygos hemiazygos veins, and thoracic duct
4. ***Nerves of the posterior mediastinum:*** vagus, splanchnic, and sympathetic chain.

Pleural cavity is the space surrounding the lungs in the thoracic cavity. There are two pleural cavities, one for each lung on the right and left sides of the mediastinum. Each pleural cavity and it's enclosed lung are lined by a serous membrane called pleura. The pleural cavity aids optimal functioning of the lugs during breathing. It transmits movements of the chest wall to the lungs, particularly during heavy breathing.

Pericardial cavity is a conical fibro-serous sac, in which the heart and the roots of the great vessels are contained. The cavity surrounds the heart and is continuous with it at all but the points of entry and exit of great vessels. It acts as mechanical protection for the heart and big vessels, and a lubrication to reduce friction between the heart and the surrounding structures.

Peritoneal cavity is the space found between the parietal and visceral layers of the peritoneum. This is filled with a small amount of serous peritoneal fluid secreted by the mesothelial cells which line the peritoneum. The peritoneal cavity 1) protects the abdominopelvic organs; 2) connects organs with each other; 3) maintains the position of organs by suspending them with ligaments; and 4) prevents friction while organs move There are two divisions of the peritoneal cavity: **lesser sac** or **omental bursa** and **greater sac** comprising the **supracolic** and **infracolic compartments**. The function of the **lesser sac** or **omental bursa** is to provide space for unhindered movement of the stomach. The supracolic compartment of the greater sac contains the liver, stomach and spleen, while the infracolic compartment contains the small intestine, ascending colon and descending colon.

Chapter 2: Homeostasis.

Homeostasis is a condition where the body resists change in order to maintain a stable, relatively constant internal environment. The stability thus attained is not static but a dynamic equilibrium, in which continuous change occurs yet relatively uniform conditions prevail. Temperature, pH, water, oxygen, blood pressure, and heart rate are just few of the internal body variables that remain within a certain range by homeostasis. When these variables fall outside the stable range, it is a condition called homeostatic imbalance which can interrupt normal body and could result in disease or illness.

Homeostasis regulation is achieved by a control system that has five basic parts: **Stimulus**, **Detector**, **Regulator**, **Effector** and **Response**.

Stimulus is a physical, chemical or environmental factor that causes deviation from the normal body environment.

A **detector** or **sensor** receives the stimulus and forwards it to the control center. Example sensory nerve endings in the skin detect pressure or temperature and move that information to the brain. We will learn more about this in chapter 5 on the integumentary system.

The **regulator** or **control center** receives and processes information from the sensor. It also sets the normal reference point for any physiological processes. Example specialized cells in the pancreas determine the normal reference point for blood glucose. We will learn more about this in chapter 6 on the endocrine system.

The **effector** is the cell, tissue or organ that responds to the signals from the regulator. : The control center commands the effector to respond to the stimulus. Example in response to increased body temperature sweat glands acts as effectors and release sweat to lower body temperature.

Response is the corrective measure taken by the effector toward the stimulus. In the above example of the sweat glands, sweat is the response. In a homeostasis control system the response can oppose or enhance the stimulus. This is achieved by the control system through feedback. There are two forms of feedback mechanisms, **negative** and **positive feedback**. Negative feedbacks counteract the change of a body property from its target value and usually dampen the effect of the initial stimuli. Positive feedback on the other hand amplifies their initiating stimuli. We learn more about positive and feedback loops in chapter 6 on the endocrine system.

A disease condition therefore is a result of the imbalance of these negative or feedback mechanisms and failure to achieve homeostasis and the previously set target point. Example **diabetes mellitus** occurs when the control mechanism for insulin becomes imbalanced, due to either insulin deficiency because the pancreas are not secreting enough insulin or because cells have become resistant to insulin like through loss of insulin receptor.

Questions to cover homeostasis

In the questions that follow there might be more than one correct answer

1. **What homeostatic responses help in balancing the body temperature during hyperthermia?**

   a. Increase the metabolic rate.
   b. Vasoconstriction to reduce blood flow to the skin.
   c. Sweating*
   d. Vasodilation to increase blood flow to the skin*.
   e. Shivering

Answer: Hyperthermia is an abnormally high body temperature. To achieve homeostasis the body would have to loose excess heat or reduce heat generating activities. Vasodilation to increase blood flow to the skin and sweating will aid in reducing the body temperature. The other options would be good choices to achieve homeostasis during hypothermia where the body temperature is abnormally low.

1. **Homeostasis regulation is achieved by a control system that has five basic parts - Stimulus, Detector, Regulator, Effector and Response. In the previous question what role did the sweat glands play?**
   a. **Stimulus**
   b. **Detector**
   c. **Regulator**
   d. **Effector**

e. Response

Answer: To achieve homeostasis during hyperthermia sweating aided in reducing the body temperature. Sweat glands that produced the sweat are effectors in this homeostatic system.

1. **What is meant by fluid balance in the body?**
   a. **Over-hydration**
   b. **Dehydration**
   c. **Sweating**
   d. **Micturition**
   e. **Water intake matches water output***

Answer: Maintaining fluid balance is a homeostatic condition and is achieved when our water intake and water output are matched.

1. **What is meant by acid-base balance in the body?**
   a. **Maintain pH at 7.4*.**
   b. **Maintain a neutral pH.**
   c. **Reducing the acidity of gastric juices**
   d. **Removal of excess acid.**
   e. **Maintain pH at 7**

Answer: In the absence of pathological states, the pH of the human body ranges between 7.35 to 7.45, with the average at 7.4. This pH is critical for many biological processes to occur. A pH below 7.35 is an acidemia, and a pH above 7.45 is an alkalemia. So acid-base balance in the body refers to the homeostatic measures taken by the body to keep the body's pH at 7.4.

1. **If the dissolved CO2 concentration in arterial blood exceeds normal levels, which of the following responses would help regain homeostasis?**
   a. **Increase in respiratory rate so that more CO2 is inhaled.**
   b. **Increase in respiratory rate so that more CO2 is exhaled*.**
   c. **Decrease in respiratory rate so that less O2 is exhaled.**

d. **Decrease in respiratory rate so that less CO2 is exhaled.**

Answer: Hypercapnia is excess carbon dioxide (CO_2) buildup in your body. Since the dissolved CO_2 in the blood is higher than normal in the stated case, in order to achieve homeostasis the levels of blood CO_2 need to be reduced. Of the options listed this is best achieved by increased respiratory rate that will increase lung ventilation and exhale more CO_2. This will lead to a decrease in the CO_2 gas dissolved in arterial blood and a return to homeostasis. Yes increased exhalation would also lead to exhaling of O_2 but the inhaled air tends to have higher O_2 and CO_2 than exhaled air.

1. **Which of the following physiological processes use a positive feedback loop?**
    a. **Acid-base balance**
    b. **Fluid balance**
    c. **Childbirth***
    d. **Thermo-regulation**
    e. **Satiety**

Answer: The release of oxytocin from the posterior pituitary gland during labor is an example of positive feedback mechanism. Oxytocin stimulates the muscle contractions that push the baby through the birth canal. The release of oxytocin result in stronger or augmented contractions during labor.

Chapter 3: The Skeletal System.

The most visible role of the skeletal system are to 1) support the body, 2) protect soft organs, and 3) generate movement. The skeletal system also has a role in 4) blood formation or hematopoiesis, 5) storing fats and 6) storing minerals. The adult human skeletal system contains 206 bones from the smallest bone the stapes - an inner ear bone measuring ~ 2.5 mm to the longest bones the femur or thigh bone measuring ~ 20 inches. In addition to bones it also consists of the joints, cartilage, tendons, and ligaments that connect them.

Types of skeletal system

The skeleton is subdivided into two major divisions—the **axial** and **appendicular** skeletal system. The **axial skeleton** is called so because it forms the central axis of the body and consists of 80 bones which form the **skull**, **vertebral column**, and **thoracic cage**.

The **appendicular skeleton** is called so because it forms the appendages of the body and connects it to the axial skeletal system. It consists of 126 bones which form the **pectoral** and **pelvic girdles**, the **limb bones**, and the bones of the **hands** and **feet**.

Types of bone

There are six types of bones in the skeleton: **flat, long, short, irregular, sutural** and **sesamoid**.

Flat bones protect internal organs such as the brain, heart and pelvic organs. They provide protection like a shield and because these bones are somewhat flattened provide large areas of attachment for muscles. Some examples of flat bones include the bones in the skull - **occipital, parietal, frontal, nasal, lacrimal,** and **vomer**; the bones in the thoracic cage - **sternum** and **ribs**; and the pelvic bones - **ilium, ischium,** and **pubis**.

Long bones are longer than they are wide and support the weight of the body and allow movement along with the muscular system. Long bones are mostly located in the **appendicular skeleton**. These include bones in the lower limbs - **tibia, fibula, femur, metatarsals,** and **phalanges**, and bones in the upper limbs - **humerus, radius, ulna, metacarpals,** and **phalanges**.

Short bones are about as long as they are wide and provide stability and some movement. They are mostly located in the **appendicular skeleton**. These include bones in the wrist also called **carpals** viz. the **scaphoid, lunate, triquetral, hamate, pisiform, capitate, trapezoid,** and **trapezium**; and the **tarsals** in the ankles - **calcaneus, talus, navicular, cuboid, lateral cuneiform, intermediate cuneiform, and medial cuneiform.**

Irregular bones vary in shape and structure and often have a fairly complex shape, which helps protect internal organs. They are located both in the **axial** and **appendicular skeleton.** The 26 irregular bones of the vertebral column in the **axial skeleton** protect the spinal cord. In the appendicular skeleton the irregular bones of the pelvis - **pubis, ilium,** and **ischium** protect the organs in the pelvic cavity.

Sutural bones are small, irregular bones within the cranial sutures. They are also called **Wormian** bones.

Sesamoid bones are small, round bones embedded in tendons and they reinforce and decrease stress on that tendon. They are located in the **appendicular skeleton** and are found in the hands, knees and feet.

Structure of a bone

The tough, thin outer membrane covering all over the bones except inside the joint space is called the **periosteum.** The periosteum consists of an outer **fibrous layer** and an inner **cambium layer.** The fibrous layer contains **fibroblasts** while the cambium layer contains **progenitor cells** which develop into **osteoblasts** that are responsible for producing more bone cells. Beneath the hard outer shell of the periosteum are tunnels and canals through which blood and lymphatic vessels run to carry nourishment for the bone.

There are **three** types of bone tissue - **compact, cancellous** and **subchondral tissue.** The **compact tissue** is the harder, outer tissue of bones. The **cancellous tissue** is sponge-like and found inside bones. The **subchondral tissue** comprises of the smooth tissue at the ends of bones. **Osteon** is the chief structural unit of **compact bone.** It consists of concentric bone layers called **lamellae,** which surround a long hollow passageway, the **Haversian canal.** The Haversian canal contains small blood vessels responsible for blood supply to osteocytes

Functions of the skeletal system

1) Support the body, 2) protect soft organs, and 3) generate movement

These are the more readily observable functions of the skeletal system. They provide a scaffold to support the body. It protects soft organs from injury by shielding them from outside force.

Bones facilitate movement by serving as points of attachment for your muscles. They serve as levers. We will learn more about levers in movement in Chapter 4 on the muscular system.

4) Blood formation or hematopoiesis, 5) storage of fats and 6) storage of minerals are the less easily observable roles of the skeletal system. The connective tissue inside most of the bones is called bone marrow and is of two types, **yellow** and **red bone marrow.**

Red bone marrow is the site of production of all blood cells including red blood cells, white blood cells, and platelets.

Yellow bone marrow contains fat or adipose tissue, and the triglycerides stored in the adipocytes of this tissue can be released to serve as a source of energy for other tissues of the body..

The skeletal system also plays a **metabolic role** where it acts as a reservoir for a number of minerals important to the functioning of the body, like **calcium**, and **phosphorus**. Based on hormonal feedback these minerals, are released back into the bloodstream when required to maintain normal physiological levels. This dissolution and resorption of the bone minerals serves as a buffer and maintains the systemic pH.

Different types of articulations or joints

Joints in the skeletal system can be classified by **structure** and **function**.

Based on structure skeletal joints can be classified as **synovial, fibrous**, and **cartilaginous**.

Synovial joints are the most common types of joints in the human body and are characterized by the presence of a fluid-filled joint cavity contained within a fibrous capsule. There are seven types of synovial joints.

1. **Ball and socket joint** – the rounded head of one bone sits within the cup of another, such as the hip joint or shoulder joint. Movement in all directions is allowed.
2. **Saddle joint** – this permits movement back and forth and from side to side, but does not allow rotation, such as the joint at the base of the thumb.

3. **Hinge joint** – the two bones open and close in one direction only (along one plane) like a door, such as the knee and elbow joints.
4. **Condyloid joint** or **ellipsoidal joint** – this permits movement without rotation, such as in the jaw or finger joints.
5. **Pivot joint** – one bone swivels around the ring formed by another bone, such as the joint between the first and second vertebrae in the neck allowing the head to turn.
6. **Gliding joint** or **plane joint** – smooth surfaces slip over one another, allowing limited movement, such as the wrist joints.
7. **Compound joint** – has more than one kind of attachment, like the knee joint where the femur and tibia have a condyloid joint and a saddle joint attaches the femur to the patella.

Fibrous joints have thick connective tissues composed of collagen fibers found between the articulations of the joints. **Cartilaginous joints** are skeletal joints that are entirely joined by cartilage, either a **fibrocartilage** or a glass -like cartilage called **hyaline cartilage**.

Based on **function** skeletal joints can be classified as **synarthrosis**, **amphiarthrosis** and **diarthrosis**.

Synarthrosis is a type of join that is strong and immovable. Examples include the skull sutures, and joint between the teeth and the mandible.

Amphiarthrosis joints allow slight movement. These include the pubic symphysis of the pelvic girdle, the cartilaginous joint that strongly unites the right and left hip bones of the pelvis.

Diarthrosis is a type of joint that allows full movement. These include joints in the elbow, shoulder, and ankle.

Ligaments are a fibrous connective tissue that attaches bone to bone, and usually serves to hold structures together and keep them stable. They are like tendons and fasciae with the difference being in what they connect. We will cover tendons and fasciae in Chapter 4 on the muscular system. Histologically, there are two types of ligaments: **white** and **yellow** ligaments.

White ligaments are rigid when the majority of the extracellular matrix is made out of **type I collagen fibers**, like the ligaments in the knee called the anterior cruciate ligament. **Yellow ligaments** are less rigid since they are mostly comprised of **elastic fibers**, like the ligaments in the spinal column called ligamentum flavum. Because of this difference yellow ligaments are more stretchy than white ligaments. This is what allows us a large range of spinal flexion without tearing the ligaments but a strong inward rotation of the knee can cause ligament tears making an anterior cruciate ligament (ACL) sprain or tear one of the most common knee injuries in sports.

Both kinds of ligaments also consist of **fibroblasts** they produce the fibers that comprise the extracellular matrix. The fibers within ligaments are densely packed and arranged in parallel, which is why ligaments are called **dense regular connective tissue**. This dense fibrillar arrangement provides maximum strength for the ligaments and increases the resistance to the stretching forces that occur during movement. In white ligaments since the main fibrillar component is composed of collagen fibers, the ligament would be more rigid and resistant to stretching. This is useful for supporting the joints and preventing the articulating surfaces from dislocating. On the other hand, if the fibers are mostly elastic like in the white ligament, it will be able to stretch more and allow wider motion range, but also during that stretching, to generate the elastic force needed to resume the primary position.

Based on tissue connectivity there are three types of ligaments: **articulation**, **peritoneal** and **periodontal** ligaments. **Articulation ligaments** are also sometimes called "true" ligaments because they connect two articulating bones as opposed to **peritoneal ligaments** which connect viscera to viscera, or viscera to the abdominal wall, or the **periodontal ligaments** which connect teeth to the alveolar bone.

The major functions of articulation ligaments are to stabilize joints and prevent excessive movement which would result in an injury.

Types of articulation ligaments

There are three types of the articulation ligaments: **capsular**, **extracapsular** and **intracapsular**. They differ by their location within a joint. To understand the nomenclature one needs to understand what is a **joint capsule**. It is a **ligamentous sac** that surrounds the articular cavity of a freely movable joint . Ligaments that form this sac/capsule are called the **capsular ligaments**, while the ligaments located outside or inside the capsule are called **extracapsular** and **intracapsular ligaments** respectively. **Capsular ligaments** enhance the strength of the articular capsule. They are present within the capsules of the joints that bear great masses, such is the hip joint. **Intracapsular ligaments**, on the other hand, are less common. They are only present in the knee, wrist, and foot. **Extracapsular ligaments** hold the entire joint in place and prevent dislocation injuries. There are two types of extracapsular ligaments: **proximate** and **remote.**

1. **Proximate extracapsular ligaments** pass over at least two joints, close to their capsules. They are found only in the leg.
2. **Remote extracapsular ligaments** are a bit more distant from the joint capsule. Most of these ligaments are yellow ligaments.

Bone Formation

Bone formation involves the transformation of a preexisting mesenchymal tissue into bone tissue. This can occur in two modes the **intramembranous ossification** and **the endochondral ossification.**

Intramembranous ossification is a process that occurs primarily in the bones of the skull and involves the direct conversion of mesenchymal tissue into bone.

In most other modes of bone formation, the mesenchymal cells differentiate into cartilage, and this cartilage is later replaced by bone. This process is called **endochondral ossification.**

Bone is composed of four different cell types; **osteoblasts, osteocytes, osteoclasts** and **bone lining cells.**

Osteoblasts are single-nucleated cells that are responsible for the synthesis and mineralization of bone during both initial bone formation and later bone remodeling. When an osteoblast is surrounded by a calcified matrix that it secreted, it is called an **osteocyte. Osteoclasts** are specialized multinucleated giant cells that resorb bone. **Bone lining cells** cover inactive or non-remodeling bone surfaces.

Epiphyseal plates are a hyaline cartilage plate containing chondrocytes or cells that secrete cartilage at the neck of the long bones on either ends. This is found only in children and adolescents whose long bones are growing. The long bones grow longitudinally as the epiphyseal plates secrete bony matrix. There are five zones in the epiphyseal plates.

Zone of reserve contains resting cartilage or chondrocytes that do not divide rapidly.

Zone of proliferation contains active chondrocytes that produce new cartilage.

Zone of maturation and hypertrophy contains chondrocytes that have stopped mitosis and enlarge due to accumulating glycogen, lipids, and alkaline phosphatase.

Zone of calcification contains chondrocytes that have started to undergo apoptosis and the cartilaginous matrix has started to calcify.

Zone of ossification is a region where the calcified cartilage is broken down and replaced with mineralized bone tissue.

In adults, who have stopped growing, the plate is replaced by an **epiphyseal line**.

Questions to cover the skeletal system

In the questions that follow there might be more than one correct answer

1. **Which bones form the lower leg**
    a. **Femur**
    b. **Hip bone**
    c. **Ulna**
    d. **Fibula***
    e. **Tibia***

Answer: Fibula and Tibia are bones that form the lower leg. Femur is a thigh bone and Ulna is one of the two bones that make up the forearm

1. **Which of the following bones are long bones?**
    a. **Radius***
    b. **Scaphoid bone**
    c. **Ulna***
    d. **Humerus***
    e. **Frontal bone**

Answer: Radius, Ulna and Humerus are long bones. The scaphoid bone is one of the carpal bones on the thumb side of the wrist and it is an irregular bone, while the frontal bone is a bone in the human skull. It is classified as a flat bone due to its relatively thin and flat shape

1. The condyles of which bone are characteristic structures of the knee joint?
    a. Femur*
    b. Radius
    c. Talus
    d. Fibula
    e. Tibia*

Answer: Femur is the upper leg bone, or thigh bone and tibia is the bone at the front of the lower leg, or shin bone. Together they form the knee joint. There is a third bone that is part of the knee joint but it wasn't listed in the choices. Patella is the thick, triangular bone that sits over the other bones at the front of the knee. It is also called the kneecap.

1. Which bones are found in the lower leg region?
    a. Humerus
    b. Fibula*
    c. Ulna
    d. Radius
    e. Tibia*

Answer: Tibia and fibula are the two long bones located in the lower leg. The tibia is a larger bone on the inside, and the fibula is a smaller bone on the outside.

1. Which long bones form the forearm (antebrachium)?
    a. Clavicle
    b. Ulna*
    c. Radius*
    d. Scapula
    e. Humerus

Answer: Radius and ulna are two long bones located in the forearm. The ulna is located on the pinky side and the radius on the thumb side.

1. **Which of the following joints permits slight movement?**
    a. **Pubic symphysis***
    b. **Transverse palatine suture**
    c. **Lambdoid suture**
    d. **Coronal suture**
    e. **Sagittal suture**

Answer: Pubic symphysis is a cartilaginous joint between the left and right superior rami of the pubis of the hip bones. This joint allows slight movement and is an example of an amphiarthrosis joint. Joints that end with the term suture are usually strong and immovable and are called a synarthrosis joint.

1. **Which bones are found in the gluteal region?**
    a. **Patella**
    b. **Fibula**
    c. **Ilium***
    d. **Tibia**
    e. **Sacrum***

Answer: The gluteal region refers to the general region of the posterior buttocks, lying external to the pelvic cavity. Ilium is the largest and uppermost bone of the hip. The sacrum is a large, triangular bone at the base of the spine that forms by the fusing of the sacral vertebrae (S1–S5). It is located in between the right and left iliac bones and connects with the ilium through the sacroiliac joints.

1. **Which long bones have epiphyses on one end only?**
    a. **Femur**
    b. **Humerus**
    c. **Radius**
    d. **Ulna**
    e. **Phalanges of hand***

Answer: Epiphyseal plates are sites on the bone where new bony matrix is secreted leading to longitudinal growth of the bone. Most long bones have the epiphyses on either end but the phalanges have epiphyses only on one side.

1. **Which bony structure lodges the spinal cord?**
   a. **Skull**
   b. **Vertebral canal***
   c. **Thoracic cage**
   d. **Peritoneal cavity**
   e. **Pleural cavity**

Answer: The vertebral canal, also known as the vertebral cavity or spinal cavity, is an anatomical space formed by the vertebral column that houses the spinal cord and the spinal nerve roots branching off the spinal cord bilaterally.

1. **The lower teeth are attached to which bone?**

   a. Maxilla
   b. Frontal bone
   c. Mandible*
   d. Temporal bone
   e. Zygomatic bone

Answer: The mandible is the largest bone in the human skull. It holds the lower teeth in place, it assists in mastication and forms the lower jawline. If you discount the bones of the inner ear then the mandible is the only movable bone of the skull. The joint between the lower teeth and the mandible is a is a fibrous mobile peg-and-socket joint called a gomphosis. Since it is a stroing and immovable joint it is also a synarthrosis type of joint.

1. **What are the cells that form the bone called?**

   a. Osteoblasts*
   b. Osteocytes
   c. Osteoclasts
   d. Bone lining cells

e. Osteoprogenitor cells

Answer: Osteoblasts are mononucleated cells that form bone. Osteocytes are mature bone cells, Osteoclasts are multinucleated cells that resorb bone. Bone-lining cells cover inactive bone surfaces. Osteoprogenitor cells are stem cells that are a precursors to an osteoblast.

1. **Which bone forms the forehead?**

a. Occipital bone
b. Parietal bone
c. Mandible
d. Maxilla
e. Frontal bone*

Answer: The frontal bone is a bowl-shaped bone in the frontal or forehead region of the skull. The primary functions of the frontal bone are the protection of the brain and the support of the structures of the head.

1. **Which joint among the following is a hinge type of joint?**

a. Shoulder joint
b. Wrist joints
c. Finger joints
d. Hip joint
e. Elbow joint*

Answer: The elbow joint is a hinge type joint and has three different portions. The ulnohumeral joint enables movement between the ulna and humerus; the radiohumeral joint allows movement between the radius and humerus; and the proximal radioulnar joint enables movement between the radius and ulna.

1. **Which of the following bones ossifies via intramembraneous ossification?**

a. Ulna
b. Humerus
c. Mandible*
d. Fibula
e. Femur

Answer: Intramembranous ossification is a mode of bone formation which involves the direct conversion of mesenchymal tissue into bone, while endochondral ossification is a process of bone formation where the cartilage is replaced by bone. Most of the bones in the body are formed by endochondral ossification while the bones of the skull are formed by intramembraneous ossification. Therefore, of the choices listed only the mandible undergoes intramembraneous ossification.

1. **Which region of the body contains only one bone?**

a. Thigh*
b. Forearm
c. Pelvis
d. Skull
e. Wrist

Answer: The femur or the thigh bone is the largest bone in the body. It is the only bone in the upper portion of the leg, and is completely covered by the thigh muscles which include the quadriceps in the front of the thigh, hamstrings in the back, gluteal muscles, and groin muscles.

1. **The human skeleton is composed of around 270 bones at birth, which is reduced to around 206 bones by adulthood after some bones get fused together. The biggest bone in the human body is the femur. Which bone is the smallest in the human body?**
   a. **Carpals**
   b. **Malleus**
   c. **Incus**
   d. **Stapes***

e. Pisiform bone

Answer: The malleus, incus, and stapes are the three smallest bones in the human body and they form the ossicular chain that connects the tympanic membrane to the oval window of the inner ear. The malleus connects to the tympanic membrane transferring auditory oscillations to the incus and then the stapes. The stapes connects to the oval window allowing for mechanical energy to be transferred to the fluid-filled inner ear. These bones play an important role in audition by amplifying and regulating the sound waves transmitted to the cochlea. Of these three, the stapes is the smallest bone in the human body.

1. **What is the name of the intercellular bone constituent which consists of ground substance, collagen fibers and inorganic mineral salts?**
    a. **Bone matrix***
    b. **Bone marrow**
    c. **Bone lacuna**
    d. **Bone canaliculus**
    e. **Periosteum**

Answer: Bone matrix also called the osteoid consists of about 33% organic matter comprising mostly of Type I collagen and ground substance, and 67% inorganic matter comprising mostly of hydroxyapatite crystals. The ground substance is the nonfibrillar organic component of the bone matrix and it fills the space around the collagen fibrils and hydroxyapatite crystals. The ground substance contains glycoproteins, proteoglycans, and glycosaminoglycans

1. **What is the name of the basic functional unit of the bone that consists of concentric lamellae of osteocytes?**
    a. **Osteoid**
    b. **Haversian canal**
    c. **Bony trabecula**
    d. **Periosteum**
    e. **Osteon***

Answer: The basic functional unit of a compact bone is the osteon. It consists of a central canal called the osteonic or haversian canal, which is surrounded by concentric rings or lamellae of bone matrix. Between the rings of matrix, the osteocytes are located in spaces called lacunae.

1. **What is the name of the double-layered connective tissue that covers the outer surface of the bone?**
    a. **Osteoid**
    b. **Haversian canal**
    c. **Bony trabecula**
    d. **Periosteum***
    e. **Lacunae**

Answer: The periosteum is a dense, fibrous connective tissue sheath that covers the bones. The outer layer, made up of collagen fibers oriented parallel to the bone, contains arteries, veins, lymphatics, and sensory nerves. The inner layer contains osteoblasts.

synovial

1. **Which cells take up the greatest portion of the inactive bone marrow?**
    a. **Adipocyte***
    b. **Megakaryocyte**
    c. **Osteoblast**
    d. **Osteocyte**
    e. **Macrophage**

Answer: Yellow bone marrow is the non-blood forming or inactive bone marrow that stores fat cells or adipocytes and triglycerides which can be released to serve as a source of energy.

Chapter 4: The Muscular System.

The main role of muscles are to 1) generate motion, 2) support body posture, and 3) heat production. Muscles attach to the skeleton by means of **tendons**, **aponeuroses**, and **connective tissue** called **fascia**. At the end of this chapter you will know the structure of the muscle, mechanism of muscle contraction, and organization of the skeletal muscle.

Muscle tissue has four characteristic properties, which are:

1. Excitability - ability to respond to stimuli;
2. Contractibility - ability to contract;
3. Extensibility - ability of a muscle to be stretched without tearing; and
4. Elasticity - ability to return to its normal shape.

A Little Bit of Biomechanics
Levers in our body are formed from bones, joints and muscles.
A lever consists of the following:

1. Bone serving as a rigid structure
2. Muscle provides a force on the bone to produce a turning movement
3. The joint serves as a fulcrum which is a fixed point
4. The body part and any additional object being moved is the load or resistance that is placed on the rigid structure

Three Types of Natural Lever Systems
In a **first class lever**, the fulcrum is in the middle of the effort and the load. This type of lever is found in the neck when raising your head up. The neck muscles provide the effort, the neck is the fulcrum, and the weight of the head is the load.

In a **second class lever**, the load is in the middle between the fulcrum and the effort. This type of lever is found in the ankle area. When standing on the tiptoed, the ball of the foot acts as the fulcrum, the weight of the body acts as the load and the effort comes from the contraction of the gastrocnemius or calf muscle.

In a **third class lever**, the effort is in the middle between the fulcrum and the load. During a triceps curl, the fulcrum is the wrist joint, the effort comes from the triceps contracting and the resistance is the weight of the palm and any weight that it may be holding.

Types of Muscle Contractions

Muscle contractions are defined by changes in the length of the muscle during contraction.

A muscle contraction in which the length of the muscle does not change is called an **isometric contraction**. Example. Pushing a wall

A muscle contraction in which the length of the muscle changes is called **isotonic contraction**. There are two types of isotonic contractions. An isotonic contraction where the muscle shortens is called a **concentric contraction**. Example: Shortening of the biceps when you raise a barbell in a bicep curl. An isotonic contraction where the muscle lengthens is called an **eccentric contraction**. Example: Lengthening of the biceps when you lower a barbell in a biceps curl.

A muscle contraction in which the velocity of the muscle contraction remains constant while the length of the muscle changes is called an **isokinetic contraction**. Example: Walking at a

fixed speed with varying force provided by the different incline angles of the treadmill.

Types of Muscles

The three types of muscle in the body are **skeletal**, **smooth**, and **cardiac** muscles.

Skeletal muscles are striated. They have cylindrical cells and multiple peripheral nuclei. Skeletal muscles are attached directly or indirectly through tendons to bones, cartilages, ligaments, fascia, or to some combination of these structures. Some are attached to organs, to skin, and to mucous membrane. These muscles produce movements of the skeleton and other body parts.

The major skeletal muscles are often called voluntary muscles because individuals can control many of them at will; however, some of their actions are automatic. For example, the diaphragm contracts automatically, but when a person is taking a deep breath they can additionally control it voluntarily, however, when taking a deep breath.

Cardiac muscles are striated and found only in the middle layer of the heart wall called myocardium. Some cardiac muscles are also present in the walls of the aorta, pulmonary vein, and superior vena cava. Cells of a cardiac muscle also have one nucleus each. The cardiac muscle consists of much broader, shorter cells that branch. Part of the boundary membranes of adjacent cardiac muscle cells make very elaborate branchings with one another. The breadth and branchings of the cardiac muscle increase its surface area for impulse conduction and form a network that facilitates coordinated contraction. These muscles have intercalated discs between the cells. Intercalated discs are part of the muscle cell membrane also called a **sarcolemma**. They contain two structures important in cardiac muscle contraction: **gap junctions** and **desmosomes**. **Gap junctions** are aggregates of intercellular channels that permit direct cell–cell transfer of ions and small molecules. In cardiac muscle fibers gap junctions allow the depolarizing current produced by cations to flow from one cardiac muscle cell to the next. This is called electric coupling, and in cardiac muscle it allows the quick transmission of action potentials and the coordinated contraction of the entire heart. This allows cardiac muscle cells to form a functional unit of contraction and it is called a **syncytium**. **Desmosomes** are intercellular junctions that tether intermediate filaments to the plasma membrane. This anchors the ends of cardiac muscle fibers together so the cells do not pull apart during the stress of individual fibers contracting. This adhesive intercellular junction is crucial to tissues that experience mechanical stress, and is also present outside the heart like in the bladder, gastrointestinal mucosa, and skin.

Smooth muscles are non-striated and are controlled involuntarily by the autonomic nervous system. These have one spindle-shaped central nucleus per cell. Smooth muscles are found in middle layer or tunica media of the wall of most blood vessels, and the muscular part of the wall of the digestive tract. They are also found in the eyeball, where it controls lens thickness and pupil size. In hollow organs that undergo peristalsis, smooth muscle cells are arranged in longitudinal and circular fashion. Contractile impulses are transmitted from one muscle cell to another at specialized sites called **nexuses**, where the plasma membranes of two smooth muscle cells are fused.

Muscles are classified histologically into **striated muscles** and **non-striated muscles** based on the structural characteristic called "striations" which is due to the arrangement of the muscle fibers actin and myosin filaments. Skeletal and cardiac muscles are grouped as striated muscles, while the visceral muscle is non-striated.

Microscopic Structure of a Skeletal Muscle

Muscle fibers are made of **myofibrils**, which are made of contractile proteins **actin** and **myosin** filaments. Actin forms the **thin filament** while myosin forms the **thick filament** in the muscle. The interaction of these two filaments creates muscle contraction. These filaments are repeated in units called **sarcomeres**, which are the basic functional units of muscle fibers. A sarcomere is broken down into a number of sections: The site of actin filament anchor is called **Z line**. The sarcomere runs in length from Z line to Z line. The site of myosin filament anchor is called **M line**. The **I band** contains only actin filaments. The **H zone** contains only myosin filaments. The length of a myosin filament, may contain overlapping actin filaments and is called an **A band**.

Muscle fibers also have **sarcoplasmic reticulum** that surrounds the myofibrils and holds a calcium ion reserve that is released during muscle contraction. It can have dilated end sacs called **terminal cisternae**, with a transverse, or **T tubule** between them. The T tubules are pathways for action potential to signal the sarcoplasmic reticulum to release calcium ions. The combination of two terminal cisternae with a T tubule is called a **triad**.

Bundles of **muscle fibers** or **fascicles** of skeletal muscles are arranged into four structural patterns, namely – **circular, parallel, convergent**, and **pennate**.

Circular skeletal muscles have a concentric ring arrangement of the fascicles. Muscles with this arrangement surround external body openings, which they close by contracting. Such muscles are called **orbicular** or **sphincter muscles**. The orbicularis muscles surrounding the eyes are an example of the circular skeletal muscles.

Convergent skeletal muscles have its fascicles converge toward a single tendon of insertion. Muscles with this arrangement are usually triangular or fan shaped. These muscles do not tend to exert as much force on their tendons. Muscle fibers can often exert opposing effects during contraction, such as not pulling in the same direction depending on the location of the muscle fiber. One example is the pectoralis major muscle found in the chest, and is responsible for flexing the upper arm.

Parallel skeletal muscles are characterized by fascicles that run parallel to one another, and contraction of these muscle groups acts as an extension of the contraction of a single muscle fiber. Most skeletal muscles in the body are parallel muscles. There are two types of parallel skeletal muscles:

1. Fusiform parallel skeletal muscles are more spindle shaped, like the biceps brachii muscle which is responsible for flexing the forearm.
2. Non-fusiform parallel skeletal muscles are more uniformly rectangular across the length of the muscle. Strap muscles have fibers that run parallel to the tendon and have a narrow belt-like appearance. They are not strong, but they do have a wide range of movement. For example the sartorius muscle of the thigh.

Pennate skeletal muscles are characterized by fascicles that are short and are attached obliquely to a central tendon that runs the length of the muscle. There are three types of pennate skeletal muscles:

1. Unipennate skeletal muscles have fascicles that insert into only one side of the tendon, like the extensor digitorum longus muscle of the leg;
2. Bipennate skeletal muscles have fascicles that insert into the tendon from opposite sides. The tendon is central giving the muscle a resemblance of a feather. The rectus femoris of the thigh is bipennate muscle;
3. Multipennate skeletal muscles look like many feathers side by side, with all their quills inserted into one large tendon. The deltoid muscle, which forms the roundness of the shoulder is an example of multipennate muscle .

How do Muscles Contract?
Sliding Filament Model

The sliding filament model describes the mechanism of skeletal muscle contraction. The key molecular components in the sliding filament model are:

Actin filaments are associated with regulatory proteins, **troponin** and **tropomyosin**.

Multiple heads of **myosin filaments** bind to sites on the actin filament.

Tropomyosin runs along the actin filament and prevents myosin attachment to actin.

Troponin binds the tropomyosin to the actin. It is made up of three parts:

Troponin I binds to the actin filament.

Troponin T binds to tropomyosin.

Troponin C can bind calcium ions.

This unique structure of troponin is the basis of **excitation-contraction coupling** in muscles.

1. Depolarization at a neuromuscular junction is conducted down the t-tubules,
2. This causes a huge influx of calcium ions into the sarcoplasm from the sarcoplasmic reticulum.
3. Calcium binds to troponin C, and moves tropomyosin away from the myosin head binding sites of the actin filaments.
4. Myosin heads bind and cross-link with the actin filaments.
5. As the myosin heads pivot they move the actin past the myosin towards the M line, this provides the power stroke.
6. ATP binding to myosin head uncouples it from actin and the filament slides back allowing the process to repeat.

Thus, the length of the filaments does not change during contraction. The length of the sarcomere decreases due to the actin filaments sliding over the myosin.

Now that we know how muscle contracts we can understand better the types of skeletal muscle fibers. There are three kinds of skeletal muscle fibers –**slow-twitch**, **fast-twitch** and **intermediate** fibers.

Slow-twitch muscle fibers also called **Type I fibers** have high concentrations of mitochondria and myoglobin, and lower glycogen content. They are smaller than fast-twitch fibers and are surrounded by more capillaries. They use aerobic metabolism to produce ATP. They produce less force and are slower to produce maximal tension due to lower myosin ATPase activity compared to fast-twitch fibers. They are fatigue resistant and are able to maintain longer-term contractions.

Fast-twitch muscle fibers also called **Type IIX** or **Type IIB** fibers have lower concentrations of mitochondria and myoglobin. They are larger than slow-twitch fibers and have lesser number of capillaries. They use anaerobic metabolism to produce ATP. They produce greater force and are quicker to produce maximal tension due to higher myosin ATPase activity compared to slow-twitch fibers. They can be easily fatigued.

Intermediate muscle fibers also called **Type IIA** are a mix of fast-twitch and slow-twitch muscle fibers. Therefore they use both aerobic and anaerobic energy systems. They fatigue more slowly than fast-twitch muscle fibers.

Tendons

A tendon is a tough, flexible band of dense connective tissue that serves to attach skeletal muscles to bones. Tendons are found at the distal and proximal ends of muscles, binding them to the periosteum of bones at their proximal (origin) and distal attachment (insertion) on the bone. As muscles contract, the tendons transmit the mechanical force to the bones, pulling them and causing movement.

Tendons have an abundance of parallel collagen fibers, which provide them with high tensile strength. Fiber bundles are called subfasicles which can then form secondary fiber bundles called fascicles. Individual fascicles are covered by a thin layer of dense connective tissue called endotenon. In turn, groups of fascicles are covered by a layer of dense irregular connective tissue called epitenon. Finally, the epitenon is encircled with a synovial sheath and attached to it by a delicate connective tissue band called mesotenon.

Fasciae

A fascia is a layer of fibrous connective tissue that attaches to, surrounds and stabilizes muscles and organs. Remember ligaments connect bone-to-bone, tendons connect muscle-to-bone while fasciae connect muscle-to-muscle. There are three layers of fascia: **superficial**, **visceral** and **deep**.

Superficial fascia is the lowermost layer of skin in nearly all regions of the body. Visceral fascia suspends organs within the body cavities. Deep fascia surrounds individual muscles.

Functions of the muscular system

As stated earlier on, the main role of muscles are to **1)** generate motion, **2)** support body posture, and **3)** heat production. Depending on the axis and plane, there are several different types of movements that can be performed by the muscular and skeletal system. Some of these movements include:

1. **Flexion** and **extension** are movements that take place within the sagittal plane and involve anterior or posterior movements of the body or limbs. These movements decrease or increase the angle between the bones involved in the movement, respectively. Example: Bending the leg at the knee joint is flexion, whereas extension would be straightening knee from a flexed position. **Hyperextension** is the abnormal or excessive extension of a joint beyond its normal range of motion, while **hyperflexion** is excessive flexion at a joint. These can result in injury.

2. **Adduction** and **abduction** are motions of the limbs, hand, fingers, or toes in the coronal plane of movement bringing the parts of the body towards or away from the midline, respectively. Example: Abduction of the arm at the shoulder joint involves moving the arm away from the side of the body, while adduction involves bringing it back towards the body.

3. **Circumduction** is a movement where the limb moves in a circle. This occurs at the shoulder joint during an overarm tennis serve.

4. **Rotation** is a movement where the limb turns round its long axis, like using a screw driver. This movement is defined relative to the midline, where **internal rotation** involves rotating the segment towards to the midline, while **external rotation** involves moving it away from the midline.

5. **Supination** and **pronation** are special types of rotatory movements used to describe the movements of the forearm or the foot. Two joints are involved in pronation and supination of the hand and forearm. These are the proximal and the distal radioulnar joints formed between the upper and lower ends of the radius and ulna, respectively. Pronation is the rotatory movement during which the palm and forearm face downward when your arms are stretched in front. Supination is the reverse movement during which the palm and forearm face upward. While considering ankle and foot movement, supination means rolling the foot outwards, while pronation means

rolling the foot inwards.

6. **Depression** and **elevation** are the downward and upward movements of the jaw or mandible when you open and close the mouth. These movements can also be seen in the shoulder blade or scapula when you shrug your shoulders.

7. **Protraction** and **retraction** are the forward and backward or anterior-posterior movements of the jaw or mandible when you jut your chin out or pull it back in. These movements can also be seen in the shoulder blade or scapula when the shoulder moves forward while throwing a ball.

8. **Excursion** is the side to side movement of the jaw or mandible. **Lateral excursion** moves the mandible either to the right or left side, away from the midline. **Medial excursion** returns the mandible to its resting position at the midline.

9. **Opposition** and **reposition** are thumb movements. The movement of the thumb that brings the tip of the thumb in contact with the tip of a finger is an **opposition** movement. It is produced by a combination of flexion and abduction of the thumb at the first carpometacarpal joint. Bringing the thumb next to the index finger involves the **reposition** movement.

The second function of the muscles is to support body posture. Muscles contribute to the overall support and stability of joints both during movement and stationary positions. Muscles and their tendons pass over joints and stabilize the articulating bones and hold them in position. Posture is maintained by a sustained tonic contraction of postural muscles. These muscles act against gravity and stabilize the body during standing or walking. The postural muscles include the muscles of the back and abdominal muscles.

The third function of the muscles is heat production. Muscle tissue is one of the most metabolically active tissues in the body, in which approximately 85 percent of the heat produced in the body is the result of muscle contraction. This makes the muscles essential for maintaining normal body temperature.

Questions to cover the muscular system

In the questions that follow there might be more than one correct answer.

1. **What structure connects a tendon to its synovial sheath?**
    a. **Endotenon**
    b. **Mesotenon***
    c. **Epimysium**
    d. **Tertiary fiber bundle**
    e. **Epitenon**

Answer: The mesotenon is a delicate sheet of connective tissue extending from the synovial membrane lining the sheath to the synovial membrane surrounding the tendon. It functions to carry the vascular supply to the tendon

1. **What is the name of the connective tissue that surrounds the whole tendon?**
    a. **Epitenon***
    b. **Tertiary fiber bundle**
    c. **Endotenon**
    d. **Mesotenon**
    e. **Epimysium**

Answer: Epitenon is the connective tissue surrounding each tendon. This allows smooth gliding of the tendon against adjacent structures.

1. **What structure do the myosin filaments form?**
    a. **I band**
    b. **Z disc**
    c. **A band***
    d. **Sarcomere**
    e. **Microfilament**

Answer: The arrangement of the thick myosin filaments across the myofibrils and the cell causes them to refract light and produce a dark band known as the A Band. In between the A bands is a light area where there are no thick myofilaments, only thin actin filaments. These are called the I Bands. Z-line also called Z-disc or Z-band bisecta the I band of striated muscle myofibrils and serves as the anchoring point of actin filaments at either end of the sarcomere.

1. **Which structure represents the area of lateral connections of thick filaments?**
   a. I band
   b. H zone
   c. Z disc
   d. M line*
   e. A band

Answer: The M line runs down the center of the sarcomere , through the middle of the myosin filaments. This serves to arrange the thick filaments into the A-bands.

1. **Rigor mortis or postmortem rigidity is a condition that appears a few hours after death. Which two structures of the sarcomere are particularly compromised in their function during rigor mortis?**
   a. H zone
   b. Sarcolemma
   c. Z disc
   d. Actin*
   e. Myosin*

Answer: The biochemical basis of rigor mortis is hydrolysis in muscle of ATP, the energy source required for movement. Without ATP, myosin molecules adhere to actin filaments and the muscles become rigid.

1. **Which of the following structures group to form the muscle bundles?**
   a. Myocyte
   b. Myosin
   c. Cardiac muscle cell
   d. Muscle fiber*
   e. Myofibril

Answer: Muscle fibers bundle together to form a muscle bundle or fasciculus. This is surrounded by a layer of connective tissue called the perimysium.

1. **Which of the following is an isokinetic contraction?**
   a. **Shortening of biceps in a barbel curl**
   b. **Lengthening of biceps in a barbel curl**
   c. **Pushing a wall**
   d. **Lying down**
   e. **Walking at a fixed speed on varying inclines of the treadmill***

Answer: Muscle contractions are defined by changes in the length of the muscle during contraction. In a barbel curl, concentric contraction of the biceps causes it to contract and eccentric contraction causes it to lengthen. Isometric contractions like while pushing a wall do not result in change in muscle length. Isokinetic contractions involve change in length of the muscle without change in velocity of the muscle contractions like while walking at a fixed speed on varying ramp inclinations on a treadmill.

1. **Which of the following is a non-striated muscle?**

   a. Smooth muscle*
   b. Skeletal muscle
   c. Cardiac muscle
   d. Myocardium
   e. Involuntary muscle

Answer: Smooth muscles are involuntary, non-striated muscles. Smooth muscle consists of thick and thin filaments that are not arranged into sarcomeres giving it a non-striated pattern. Compared to skeletal muscles they are contracted and controlled involuntarily. Cardiac muscles are also involuntary but they are also striated so selecting involuntary muscle is not a correct option for this question.

1. **What kind of movement does the arm undergo at the shoulder joint when it moves away from the side of the body?**

   a. Adduction*
   b. Abduction
   c. Protraction
   d. Retraction
   e. Excursion

Answer: Adduction is motion of the limbs, hand, fingers, or toes that brings the parts of the body away from the midline.

1. **What molecule prevents myosin attachment to the actin filament?**

   a. Troponin C
   b. Troponin I
   c. Troponin T
   d. Tropomyosin
   e. ATP

Answer: Tropomyosin runs along the actin filament and prevents myosin attachment to actin.

Chapter 5: The Integumentary System.

The integumentary system consists of the skin and the associated structures like glands and hair. It is the largest organ in the body weighing about 9lbs and covering an area of about 20 square feet in an adult. It provides a tough, waterproof layer that protects the insides of the body. The sensory receptors in the skin send information to the nervous system and lets us appreciate the texture and temperature of our environment; it regulates body temperature; it allows excretion through sweating. The skin lets in small amount of UV light into the subcutaneous fat that helps in vitamin D production. Hair on our head and body hairs keep us warm. The nails protect fingers and toes. The skin also contains immune system cells and the acidic surface of our skins retards growth of most pathogens.

The integumentary system is composed of the following parts:

1. Skin
2. Skin appendages
    ◦ Hairs
    ◦ Nails
    ◦ Sweat glands
    ◦ Sebaceous glands
3. Subcutaneous tissue and deep fascia
4. Mucocutaneous junctions

Structure of the skin

Skin can be thin, hairy or glabrous. Glabrous skin is free from hair and is found over the palms, soles of the feet and flexor surfaces of the fingers. The skin is composed of three layers; the **epidermis**, **dermis** and **hypodermis**.

Epidermis

The epidermis is the topmost layer of the skin, and is largely formed by layers of keratinocytes undergoing terminal maturation. Non-keratinocyte cells in epidermis include:

Melanocytes that produce melanin and pigment formation. **Langerhans cells** that are antigen-presenting cells of the immune system. **Merkel cells** that are the sensory mechanoreceptors.

The epidermis can be divided into the following **five** layers from the deepest to most superficial:

Stratum basale is the innermost layer and it has keratinocytes undergoing mitosis in this layer. Pigment producing melanocytes are in this layer. **Stratum spinosum** contains cells called prickle cells that have small radiating processes that connect with other cells. Keratin is synthesized in this layer. **Stratum granulosum** contains cells that secrete lipids and other waterproofing molecules. **Stratum lucidum** cells have lost their nuclei and have increased keratin production. **Stratum corneum** is the outermost layer of cells that have lost all organelles and have been hardened with keratin.

Dermis

The middle layer of the integument is known as the dermis. The dermis has two layers, the **superficial papillary layer**, and the deeper **reticular layer**. The papillary layer is responsible for fingerprints. The reticular layer has thicker bundles of collagen fibers. There are a number of accessory structures of the skin in the dermis. These include the following:

Fibroblasts are cells that synthesize the extracellular matrix, composed of collagen and elastin. **Mast cells** are cells of the immune system that contain histamine granules which are released during an allergic reaction. The dermis also houses **blood vessels, cutaneous sensory nerves, hair follicles, nails, sebaceous** and **sweat glands** also called **glandulae sudoriferae**. There are two main types of sweat glands, the **eccrine** and the **apocrine** glands. The **eccrine glands** are the major sweat glands of the human body that release sweat that is mostly sodium chloride and water which aids in temperature control. **Apocrine glands** are larger sweat glands, located in the axillary and genital regions. The sweat from apocrine glands can be broken down by cutaneous microbes, producing body odor.

The sebaceous glands occur closely with the hair follicles and are together called a **pilosebaceous unit**. The normal function of sebaceous glands is to produce and secrete sebum, a group of complex oils including triglycerides and fatty acid breakdown products, wax esters, squalene, cholesterol esters and cholesterol. Sebum lubricates the skin to protect against friction and makes it more impervious to moisture. In addition to the sebum gland the hair follicle has the following parts: the **follicle** consisting of the **infundibulum**, the **isthmus** and the **bulb**. The other structures include the **dermal papilla** and the **arrector pili muscle**. Contraction of the arrector pili causes the follicle to stand upright.

Hairs

Hairs are filamentous cornified structures which grow out of the skin and cover most of the body surface. Several areas of the body like the palms, and soles etc. are devoid of hairs. Hairs are important in sensing, thermoregulation and protection against injury and solar radiation. There are two types of hairs: **vellus** and **terminal**.

Vellus hairs do not project beyond their follicles in some of the areas, however, they are short and narrow and cover most of the surface of the body. This hair type is seen on children. **Terminal hairs** are longer, thicker and more heavily pigmented. They are mostly observed on males but also in the pubic regions of both sexes.

Strands of hair originate in an epidermal penetration of the dermis called the **hair follicle**. The **hair shaft** is the keratinized part of the hair not anchored to the follicle, and much of this is exposed at the skin's surface. The rest of the hair, which is anchored in the follicle, lies below the surface of the skin and is referred to as the **hair root**. The hair root ends deep in the dermis at the hair bulb, and includes a layer of mitotically active basal cells called the **hair matrix**. The **hair bulb** is the lowest expanded extremity of the hair follicle that fits like a cap over the dermal **hair papilla**, enclosing it. The dermal **hair papilla** is a cluster of mesenchymal cells giving rise to several capillaries, which form a capillary loop. The hair bulb generates the hair and its inner root sheath.

Nails

Nails are a horny plate homologous to the stratum corneum of the epidermis that grows on the back of fingers and toes at its end. They consist of compacted and layered keratin-filled squames. The arrangement and cohesion of the squames are responsible for the hardness of nails. A nail consists of: the **nail plate, nail folds, nail matrix, nail bed** and **hyponychium**.

Nail plate: The nail plate is the outer portion of of the nail formed by layers of keratin. It forms a convex, translucent plate. It originates from the nail matrices, and is found at the base of the nails.

Nail folds: The nail folds are skin that surrounds and protects the borders of the nail plate, laterally and proximally. The **cuticle** or **eponychium** is a layer of stratum corneum which extends between the skin of the finger and the proximal nail plate.

Nail matrix: The nail matrix also called germinal matrix is the structure out of which the nail plate grows. Cells within this layers become keratinized to form the nail plate.

Nail bed: The nail bed lies underneath the nail plate. It provides a smooth surface for the growing nail to slide over. The distal margin of the nail bed is called the **onychodermal band**. There is a perfect match between the nail bed and plate, forming a seal, which prevents microbial invasion and debris collection.

Hyponychium: The hyponychium is the skin under the free edge of the nail. It is located beyond the distal end of the nail bed, near the fingertip. As a barrier from germs and debris, the hyponychium stops external substances from getting under the nail.

Hypodermis

The **hypodermis**, or **subcutaneous tissue**, or **superficial fascia** lies below the dermis. It is a major body store of adipose tissue. It provides both structural support to the skin and acts as an insulator from cold and aids in shock absorption.

Mucocutaneous junctions

These are regions of the body where there is a transition from mucosa to skin. At such regions, epithelium transitions to epidermis, lamina propria changes to dermis and smooth muscle becomes skeletal muscle. They occur at orifices in areas like the lips, nostrils, conjunctivae etc.

Blood supply to the Integumentary System

The integumentary system is supplied by the cutaneous circulation, which is crucial for thermoregulation. It consists of three types: **direct cutaneous, musculocutaneous** and **fasciocutaneous** systems. The **direct cutaneous circulation** are derived directly from the main arterial trunks and drain into the main venous vessels. **Musculocutaneous vessels** arise from intramuscular vasculature after piercing muscles and spreading out in the subcutaneous tissue. **Fasciocutaneous blood vessels** consist of perforating branches from vessels located deep to the deep fascia.

Nerve Innervation of the Integumentary System

The largest part of the innervation of the integumentary system is for the skin to facilitate its great sensorial capabilities. These include **Pacinian corpuscles**, **Meissner's corpuscles** and a large variety of other receptors for a range of stimuli.

Pacinian corpuscles are onion-like structures in the dermis and hypodermis. They contain a myelinated nerve ending in the central core of the structure. The outer layers are composed of flattened cells, collagen fibers and a lymph-like fluid. Pacinian corpuscles are sensitive to mechanical and vibratory pressure.

Meissner's corpuscles are localized in the dermis between epidermal ridges. They contain an unmyelinated nerve ending surrounded by Schwann cells. Meissner's corpuscles are touch receptors and enriched in fingers and toes.

Merkel's disks are a slowly adapting cutaneous mechanoreceptors. These receptors respond to indentation of the skin. They adapt slowly to pressure, and therefore record the sustained presence of pressure on the skin.

Ruffini's corpuscles are another class of slowly adapting cutaneous mechanoreceptors. They respond to sustained pressure and show very little adaptation. Ruffinian endings are located in the deep layers of the skin where they register mechanical deformation within joints as well as continuous pressure states.

The components of the integumentary system receive their innervation, mostly autonomic, via spinal and cranial nerves. The nerve endings branch out and form reticular plexuses in the dermis, innervating the respective components.

Questions to cover the Integumentary System

In the questions that follow there might be more than one correct answer

1. **Which of the following glands can only be found in hairy skin and are usually associated with hair follicles?**
   a. **Lacrimal gland**
   b. **Eccrine gland**
   c. **Apocrine gland**
   d. **Sebaceous gland***
   e. **Submucosal gland**

Answer: Sebaceous glands in the skin are almost always associated with hair follicles. They secrete an oily substance called sebum that lubricates and waterproofs the skin.

1. **Which layer of the skin contains pigment containing cells?**
    a. **Papillary layer**
    b. **Stratum corneum**
    c. **Stratum basale***
    d. **Stratum lucidum**
    e. **Stratum granulosum**

Answer: The stratum basale is the deepest epidermal layer and attaches the epidermis to the basal lamina. This layer also has melanocytes.

1. **Which layer of the skin gives rise to new cells?**

a. Papillary layer
b. Stratum corneum
c. Stratum basale*
d. Stratum lucidum
e. Stratum granulosum

Answer: The stratum basale is also called the stratum germinativum due to the fact that it is constantly germinating new cells.

1. **Which epidermal layer lies above the stratum lucidum?**

a. Papillary layer
b. Stratum corneum*
c. Stratum basale
d. Stratum spinosum
e. Stratum granulosum

Answer: Stratum corneum is the outermost layer of cells that have lost all organelles and have been hardened with keratin. They lie above the stratum lucidum.

1. **What is the name of the openings in the epidermis, through which the secretion of sweat glandulae sudoriferae is released onto the surface of the skin?**

a. Hair follicle
b. Sweat pore*
c. Sebaceous gland
d. Pacinian corpuscle
e. Gap junction

Answer: Glandulae sudoriferae are also called sweat glands. Their secretions are released through the sweat pores.

1. **What is the name of the keratinized part of the hair?**

a. Hair follicle
b. Hair root
c. Dermal papilla
d. Hair shaft*
e. Hair bulb

Answer: Strands of hair originate in the **hair follicle**. The **hair shaft** is the keratinized part of the hair not anchored to the follicle. The **hair root** is portion of the hair below the surface of the skin that ends in a structure called the **hair bulb**..

1. **Which structure mainly consists of loose connective tissue and adipose tissue and serves the human body for insulation?**

a. Epidermis
b. Stratum lucida
c. Fascia
d. Dermis
e. Subcutaneous tissue*

Answer: The subcutaneous tissue, also known as hypodermis or superficial fascia lies below the dermis and is a major body store of adipose tissue that serves for insulation.

1. **What are the components of a pilosebaceous unit? structure is always associated with the arrector pili muscle and a sebaceous gland?**

   a. Melanocyte
   b. Sebaceous gland*
   c. Hair follicle*
   d. Hair shaft*
   e. Pacinian corpuscle

Answer: The hair follicle, hair shaft and sebaceous gland are collectively known as the pilosebaceous unit and it originates from the basal layer of the epidermis.

1. **Which of the following layers of the skin is responsible for the fingerprints?**

   a. Reticular layer
   b. Stratum corneum
   c. Stratum lucidum
   d. Papillary layer
   e. Stratum basale

Answer: The papillary layer of the dermis is responsible for fingerprints.

1. **Which structure is located in the dermal papillae and enables the perception of touch?**

   a. Meissner's corpuscles*
   b. Tactile corpuscles*
   c. Pacinian corpuscle
   d. Ruffini's corpuscles

e. Hair follicle

Answer: Meissner's corpuscles, also known as tactile corpuscles, are found in the upper dermis, and they project into the epidermis. They are found primarily in the glabrous skin on the fingertips and eyelids. They respond to fine touch and pressure, but they also respond to low-frequency vibration or flutter.

Chapter 6: The Endocrine System.

The endocrine system is a collection of glands that release hormones into the bloodstream. These hormones help control mood, growth and development, the function of our organs, metabolism , and reproduction. There is a system of negative and positive feedback that regulates how much of each hormone is released.

While studying the endocrine system it is important to take note of the name, location and secretion of each endocrine organ; and target tissue, and effect of the hormones.

Hormones are chemicals secreted by a gland. They usually affect distant target tissue. Hormones vary in their range of targets. Some types of hormones can bind with compatible receptors found in many different cells all over the body. Other hormones are more specific, targeting only one or a few tissues. Additionally the same tissue can act as a target for many different hormones.

There are two classifications for hormones, either based on how they travel within the body or based on their chemical structure.

Autocrine, paracrine and **synaptic** are three types of local hormone signaling. **Autocrine signaling** is a situation when hormones from the same cells affect other process within the secreting cells. In **paracrine signaling**, hormones are released into the fluid between cells and diffuse to nearby target cells. **Synaptic signaling** is a more specialized signaling that occurs between neurons and between neurons and muscle cells. , allowing nerve cells to talk to each other and to muscles.

The many different hormones in the human body, they can be divided into three classes based on their chemical structure: **lipid-derived, fatty acid-derived, amino acid-derived,** and **peptide** and **protein** hormones. Lipid-derived hormones are mostly derived from cholesterol. If a hormone is amino acid-derived, its chemical name will usually end in –ine. Lipid-derived hormones can diffuse across plasma membranes whereas the amino acid-derived and peptide hormones cannot. **Group I hormones** bind to intracellular receptors and are lipid derived, while **group II hormones** bind to cell surface receptors and are hydrophilic.

Regulation in endocrine signaling can be through a negative feedback or positive feedback. A negative feedback is a sequence of biochemical events triggered by a hormone that results in formation of a substance that inhibits the formation of its precursors. When the sequence of biochemical events results in a product that increases the formation of its precursors it is called a positive feedback.

Endocrine glands and their secretions

Hypothalamus Gland

The hypothalamus is composed mainly of different nuclei or discrete masses of grey matter in the central nervous system, that synthesize different hormones in response to physiological changes. These nuclei have been grouped into four regions -the **preoptic region**, **supraoptic region**, **tuberal region** and the **mammillary region**. Anterior and posterior branches of the **circle of Willis** provide arterial blood to the hypothalamus. The hypothalamus also receives arterial supply from the hypothalamic branches of the **superior hypophyseal artery**. The hypothalamic arteries have anastomotic connections to the primary and secondary capillary plexuses of the pituitary gland. The infundibulum is a tube-like structure that connects the posterior pituitary to the hypothalamus. It allows for hormones synthesized in the hypothalamus to be sent to the posterior pituitary for release into the bloodstream.

Some of the hormones secreted by the **hypothalamus** are:

1. **TRH** or **thyroliberin** or **thyrotropin-releasing hormone** tells the pituitary to release thyrotropin.
2. **CRH** or **corticoliberin** or **corticotropin-releasing hormone** tells the pituitary to release corticotropin.
3. **GnRH** or **gonadoliberin** or **gonadotropin-releasing hormone** tells the pituitary to release gonadotropin.
4. **GHRH** or **somatoliberin** or **growth hormone-releasing hormone** tells the pituitary to release somatotropin.
5. **Follistatin** tells the pituitary to inhibit follicle-stimulating hormone.

Pituitary Gland

The **pituitary gland** or the **hypophysis** is a pea-sized gland that sits within a small depression in the **sphenoid bone**, known as the **sella turcica**. The pituitary is divided into the **anterior lobe** or **adenohypophysis** and a **posterior lobe** or **neurohypophysis**. The anterior lobe can be divided into three parts, the **Pars anterior, Pars intermedia** and **Pars tuberalis**.

The **adenohypophysis** receives blood supply from the superior hypophyseal artery while the neurohypophysis receives blood through the **superior hypophyseal, infundibular** and **inferior hypophyseal arteries**. The **anterior** and **posterior hypophyseal veins** provide venous drainage to both parts of the pituitary gland. The pituitary gland is also sometimes called the "master" gland because its hormones control many other endocrine glands.

The **adenohypophysis** secretes the following **six** hormones

1. **HGH** or **Human Growth Hormone** targets bones and soft tissue. It accelerates rate of body growth.
2. **TSH** or **Thyroid Stimulating Hormone** targets the thyroid gland. It stimulates synthesis and release of thyroid hormones.
3. **ACTH** or **Adenocorticotropic Hormone** targets the adrenal cortex. It stimulates release of glucocorticoids.
4. **PRL** or **Prolactin** targets the mammary glands. In females it promotes development of mammary glands and stimulates milk production.
5. **FSH** or **Follicle Stimulating Hormone** targets the ovaries and testes. In females it promotes growth of ovarian follicles and in males it stimulates spermatogenesis.
6. **LH** or **Luteinizing Hormone** targets the ovaries and testes. In females it promotes growth of ovarian follicles, ovulation and secretion of estrogen and progesterone. In males it stimulates testosterone secretion.

The **neurohypophysis** secretes the following **two** hormones.

1. **ADH** or **Antidiuretic Hormone** targets the kidneys. It promotes reabsorption of water in the distal convoluted tubules and collecting ducts of the kidney.
2. **Oxytocin** targets the mammary glands and uterus. In females it promotes secretion of milk and contraction of the uterus.

Thyroid Gland

The **thyroid gland** is a butterfly shaped endocrine gland located in the anterior neck. It has two lobes, left and right which are connected by a central isthmus and are wrapped around the cricoid cartilage and superior rings of the trachea.

The thyroid gland receives blood through two arteries - the **superior** and **inferior thyroid arteries**. The venous drainage is carried by the **superior, middle**, and **inferior thyroid veins**, which form a **venous plexus** around the thyroid gland.

The **thyroid** secretes the following **three** hormones.

1. **T3** or **Triodothyronine** acts on numerous target tissues, such as the brain, bone, heart and the muscles. It accelerates metabolic rate, oxygen consumption, and glucose consumption.
2. **T4** or **Tetraiodothyronine** like T3 also acts on numerous target tissues, such as the brain, bone, heart and the muscles. It accelerates metabolic rate, oxygen consumption, and glucose consumption.
3. **Calcitonin** is produced by the C cells of the thyroid gland and is released when levels of calcium in the blood are increased. It decreases the amount of calcium in the blood.

Because of two major hormones produced by thyroid gland having iodine in it, the thyroid gland is sensitive to iodine deficiencies. Goiter is a condition in which the thyroid gland becomes larger either due to inflammation or due to iodine deficiency.

Parathyroid Gland

The **parathyroid glands** are flattened and oval in shape, embedded on the posterior surface of the thyroid gland. It consists of the **superior** and **inferior parathyroid glands**. The vascular supply is of the parathyroid gland is similar to that of the thyroid gland. It receives blood through two arteries - the **superior** and **inferior thyroid arteries**. The venous drainage is carried by the **superior, middle**, and **inferior thyroid veins**.

The **parathyroid** secretes the **PTH** or **Parathyroid Hormone** which acts on the kidneys, skeletal system, and intestine. It regulates the amounts of calcium, phosphorus and magnesium in the bones and blood.

Adrenal Gland

The **adrenal** (or **suprarenal**) glands are located in the posterior abdomen on the upper poles of each kidney. The right gland is **pyramidal** in shape, while the left is **semi-lunar**. The adrenal glands consist of an outer connective tissue **capsule**, a **cortex** and a **medulla**. The **cortex** and **medulla** are the functional portions of the gland. The cortex can be divided into three regions - the **Zona glomerulosa, Zona fasciculata,** and **Zona reticularis.** The medulla lies in the center of the adrenal glands and has **chromaffin cells.** The adrenal glands are supplied blood through the **superior, middle** and **inferior adrenal arteries.** Venous drainage of the glands is through the **right** and **left adrenal veins.**

The **adrenal cortex** secretes the following **three classes** of hormones.

1. **Mineralocorticoids,** like **aldosterone** which targets the kidneys, colon, sweat and salivary glands. It helps to maintain the body's salt and water levels which, in turn, regulates blood pressure.
2. **Glucocorticoids,** like **cortisol** targets the liver, muscle, adipose tissue, and pancreas. It stimulates glucose production, has anti-inflammatory effects, promotes vasoconstriction and helps the body resist stress.
3. **Adrenal androgens** like **DHEA** or **dehydroepiandrosterone** which is a precursor hormone. It has little biological effect on its own, but when converted into other hormones such as **testosterone** and **estradiol** it has powerful effects on the body.

The **adrenal medulla** secretes the **amine hormones** like **epinephrine** and **norepinephrine** which targets most cells in the body. It increases metabolism, heart rate and output in response to stress and exercise.

The mixed glands of the endocrine system are glands with both endocrine and exocrine function. Some of the major mixed glands and their secretions are:

Pancreas

The **pancreas** is an abdominal glandular organ with both digestive and hormonal functions. We will discuss digestive functions of the pancreas in chapter 7. The part of the pancreas that plays an endocrine role consist of scattered groups of cells called pancreatic islets or islets of Langerhans. The islets consist of four distinct cell types, of which three (alpha, beta, and delta cells) produce hormones. The fourth component, the C cells do not secrete hormones. The pancreas are supplied blood by the **superior** and **inferior pancreaticoduodenal arteries**. The head of the pancreas drain blood into the **hepatic portal vein**. The rest of the pancreas drain via the **splenic vein**.

The **islets of Langerhans** secretes the following **three** hormones.

1. **Glucagon** is secreted by the alpha cells which targets the liver and muscles. It stimulates glycogenolysis, a process by which glycogen, the primary carbohydrate stored in the liver and muscle cells, is broken down into glucose to provide energy and to maintain blood glucose levels during fasting.

2. **Insulin** is secreted by the beta cells which targets the liver, the skeletal muscle and the adipose tissue. It stimulates movement of blood glucose across cells, and glycolysis, a process by which glucose is broken down to release energy and pyruvic acid. This in turn reduces blood glucose levels.

3. **Somatostatin** is secreted by the delta cells. It targets several areas of the body. In the hypothalamus, it regulates the secretion of pituitary gland hormones. In the pancreas, it inhibits the secretion of glucagon and insulin. In the gastrointestinal tract, it reduces gastric secretion and the emission of gastrointestinal hormones, such as secretin and gastrin.

Thymus

The **thymus gland** is located behind the sternum, between the lungs and has a lobular structure with the follicles having a **medullary** and **cortical portion**. The thymus gland regresses in adulthood and it shrinks and is replaced with fat. The blood supply to the thymus gland is through the **anterior intercostal** and **internal thoracic arteries**. Venous blood drainage is through the **left brachiocephalic** and **internal thoracic veins**.

The **thymus** secretes **thymosin** which stimulates the development of T cells. Thymus also has lymphoid functions since T-cells mature in the thymus.

Pineal gland

The pineal gland is small pine cone shaped glandular body in the brain. It has two types of cells, the **pinealocytes** and **glial cells**. In middle age, the gland becomes calcified. The blood supply to the pineal gland is through the **posterior choroidal arteries** while the venous blood drainage is through the **internal cerebral veins**.

The pineal gland secretes **melatonin** which targets the **suprachiasmatic nuclei** in the **hypothalamus** of the brain. It regulates the sleep cycle or circadian rhythm and the release of reproductive hormones.

Questions to cover the endocrine system

In the questions that follow there might be more than one correct answer

1. **Which endocrine gland is considered a "Master gland" that regulates other endocrine glands?**
    a. **Pituitary Gland***
    b. **Pancreas**
    c. **Thymus**
    d. **Parotid Gland**
    e. **Hypophysis***

Answer: The pituitary gland or the hypophysis is a pea-sized gland whose secretions control many other endocrine glands and therefore it is also called a "master" gland.

1. **Which endocrine organ requires iodine to maintain its proper function and it is deficiency enlarges to form goiter?**
    a. **Thymus**
    b. **Pineal Gland**
    c. **Thyroid Gland***
    d. **Suprarenal Gland**
    e. **Pancreas**

Answer: A goiter commonly develops as a result of iodine deficiency or inflammation of the thyroid gland.

1. **Which of the following glands is present in the brain?**

a. **Thymus**
b. **Pineal Gland***
c. **Thyroid Gland**
d. **Suprarenal Gland**
e. **Pancreas**

Answer: The pineal gland is small pine cone shaped glandular body in the brain that secretes melatonin and regulates the sleep cycle or circadian rhythm.

1. **Which of the following glands is both an endocrine organ as well as a lymphoid organ?**

a. Thymus*
b. Pineal Gland
c. Thyroid Gland
d. Suprarenal Gland
e. Pancreas

Answer: Thymus is both an endocrine organ and a lymphoid organ. It's endocrine role is in secreting thymosin while its lymphoid role is in being the site of T cell maturation.

1. **Which of the following glands secretes both enzymes and hormones as part of its function?**

a. Thymus
b. Pineal Gland
c. Parotid Gland
d. Hypophysis
e. Pancreas*

Answer: The pancreas is an abdominal glandular organ with both digestive and hormonal functions. As part of its digestive function it secretes enzymes like lipases, proteases and amylases. As part of its hormonal functions it secretes insulin, glucagon, and somatostatin.

1. **Defect in which gland causes diabetes insipidus?**

a. Thymus
b. Pineal Gland
c. Parotid Gland
d. Hypophysis*
e. Pancreas

Answer: If you picked pancreas you were thinking of a condition called diabetes mellitus. Diabetes insipidus is a condition which causes the body to make too much urine. The patient tends to be dehydrated and thirsty because of this. Unlike in diabetes mellitus, the blood glucose levels are normal in diabetes insipidus. It can be caused due to defects in ADH or Antidiuretic hormone secretion from the pituitary or hypophysis. ADH targets the kidneys. It promotes reabsorption of water in the distal convoluted tubules and collecting ducts of the kidney.

1. **Which gland is atrophied in adults?**

a. Thymus*
b. Pineal Gland
c. Parotid Gland
d. Hypophysis
e. Pancreas

Answer: The thymus gland regresses in adulthood and it shrinks and is replaced with fat.

1. **Which of the glands have a duct?**

a. Thymus
b. Pineal Gland
c. Sebaceous Gland*
d. Hypophysis

e. Pancreas*

Answer: Endocrine glands tend to be vascular and do not have ducts while exocrine glands tend to have ducts. Sebaceous glands secrete sebum and are an exocrine gland. Pancreas have both endocrine and exocrine functions. It has the duct of Wirsung, the major pancreatic duct that joins the pancreas with the bile duct.

1. **The Zona glomerulosa, Zona fasciculata, and Zona reticularis are parts of which gland?**

   a. Pineal Gland
   b. Hypothalamus
   c. Adrenal Gland*
   d. Hypophysis
   e. Thyroid Gland

Answer: The cortex of the adrenal glands can be divided into three regions - the **Zona glomerulosa, Zona fasciculata,** and **Zona reticularis.**

1. **The infundibulum connects which endocrine glands?**

   a. Pineal Gland
   b. Hypothalamus*
   c. Parathyroid Gland
   d. Hypophysis*
   e. Thyroid Gland

Answer: Infundibulum hypothalamus also known as pituitary stalk or infundibulum connects the posterior pituitary to the hypothalamus.

Chapter 7: The Digestive System.

The next time you sit down to enjoy your burger take a moment to think how we are what we eat. Your burger is chewed up, swallowed and moves down a long and winding tube to become part of your body.

The two basic processes of digestion are 1) the chemical and mechanical breakdown of the food; and 2) the absorption of the constituents of the food.

The digestive system at its simplest is a long tube called the alimentary canal, beginning at the mouth and ending at the anus. The length of the alimentary canal ranges approximately 23 feet long in adults. The canal has the same number of layers but different tissue characteristics throughout its length. The layers are the **mucosa** which is the inner mucus membrane, followed by the **submucosa** that contains blood and lymphatic vessels. There are two layers of smooth muscles. One layer runs lengthwise while the other encircles the tube. This arrangement of the smooth muscles aids in the **peristaltic** movement of food through the alimentary canal. The serosa is the outer serous membrane that secretes a slimy, serous fluid.

Let's now talk of the key anatomical structures in the digestive system.

Mouth is the site of ingestion and breaking down of the food. Digestion of carbohydrates and fats starts in the mouth because of the enzymes in the saliva. The mouth is surrounded by a **roof**, a **floor**, and the **cheeks**. The tongue fills a large proportion of the cavity of the mouth. The roof of the mouth consists of the **hard** and **soft palates**. The cheeks are formed by the **buccinator muscle**. The floor of the mouth consists of a **muscular diaphragm, geniohyoid muscles** and the **tongue**. The buccinator muscle contracts to keep food between the teeth when chewing. The muscular diaphragm is made of the **bilateral mylohyoid muscles** and it provides structural support to the floor of the mouth. It also pulls the larynx forward during swallowing. The geniohyoid muscles are involved in pulling the larynx forward during swallowing. The tongue is involved in moving the food inside the mouth and is connected to the floor by the frenulum of the tongue, a fold of oral mucosa. The tongue helps in gustation, or the act of tasting through its taste buds that allow us to detect sweet, sour, salty, bitter, and umami basic tastes.

Vasculature of the mouth

The mouth receives blood from various branches of the **external carotid artery**. The **superior labial** branches of the **facial** and **infraorbital** arteries supply blood to the upper lip while the **inferior branches** of the **facial** and **mental** arteries supply the lower lip. The **superior alveolar** arteries from the **maxillary artery** supply blood to the upper teeth while the **inferior alveolar arteries** supply blood to the lower teeth. The **greater** and **lesser palatine** arteries supply the palate. The tongue gets its blood supply from the **branches** of the **lingual artery**. The **dorsal lingual arteries** supply the posterior part of the tongue, the **deep lingual artery** supplies the anterior part of the tongue and **sublingual artery** supplies the sublingual gland and the floor of the mouth. The veins of the mouth generally follow the arteries and have the same names. The veins of the palate drain into the **pterygoid venous plexus**. The **lingual** veins of the tongue drain into the **internal jugular vein**.

Esophagus is a fibromuscular tube that transports food from the **pharynx** to the **stomach**. The **upper** and **lower esophageal sphincters** prevent the entry of air and the reflux of gastric contents respectively.

Vasculature of the esophagus

The blood vasculature to the esophagus can be divided into its thoracic and abdominal components. The thoracic part of the esophagus receives blood from the branches of the **thoracic aorta** and the **inferior thyroid artery**. While the venous drainage occurs via branches of the **azygous** veins and the **inferior thyroid** vein. The abdominal part of the esophagus is supplied by the **left gastric** artery and **left inferior phrenic artery**. While the venous drainage occurs via the **left gastric** vein to the **portal circulation** and the **azygous** vein to the **systemic circulation**.

Stomach is the major site of chemical digestion of proteins and some fats, but very little absorption of the digested food happens here. The stomach has four main anatomical divisions called the **cardia, fundus, body** and **pylorus**. The cardia surrounds the opening of the stomach. The **inferior esophageal sphincter** marks the transition point between the esophagus and stomach. The fundus is often a gas filled portion superior to and left of the cardia. The body is the central portion of the stomach. The pylorus connects the stomach to the duodenum and is divided into the **pyloric antrum, pyloric canal** and **pyloric sphincter**. The pyloric sphincter controls the exit of food and gastric acid mixture (called chyme) from the stomach.

Vasculature in the stomach

The arterial supply to the stomach comes from the branches of **celiac trunk** the **right gastric** and **left gastric** arteries and the **right** and **left gastro-omental** arteries. The right gastro-omental artery is the terminal branch of the gastroduodenal artery, which arises from the common hepatic artery. The left gastro-omental is a branch of the splenic artery, which arises from the coeliac trunk. The veins of the stomach run parallel to the arteries. The **right** and **left gastric veins** drain into the **hepatic portal vein**. The **short gastric vein, left** and **right gastro-omental veins** drain into the **superior mesenteric vein**.

Duodenum , a part of the small intestine is where the fats and the remaining proteins get digested but most of the digested food is not absorbed here.

The duodenum can be divided into four parts: **superior, descending, inferior** and **ascending**. Together these parts form a 'C' shape, that wraps around the head of the pancreas.

Jejunum and **Ileum** are the distal two parts of the small intestine where the last of the food gets digested and most of the food gets absorbed.

Vasculature in the duodenum, jejunum and ileum

The duodenum is supplied blood by the **gastroduodenal** artery, which is a branch of the common hepatic artery from the coeliac trunk; and the **inferior pancreaticoduodenal** artery, which is a branch of superior mesenteric artery. The veins of the duodenum follow the major arteries and drain into the hepatic portal vein. The arterial supply to the jejunoileum is from the **superior mesenteric** artery while the venous drainage is via the **superior mesenteric** vein which joins with the splenic vein at the neck of the pancreas to form the hepatic portal vein.

The **colon**, also called the large intestine is the major site for extracting water and electrolytes from the undigested remains of the food and form feces. The colon can be divided into four parts (proximal to distal): **ascending, transverse, descending** and **sigmoid**. The **sigmoid colon** is a site of storage of gases or flatus that is made in the stomach and intestines as the body breaks down food. At its distal end the sigmoid colon connects to the **rectum**.

There are a number of characteristic features that allow to distinguish the large intestine from the small intestine. The large intestine has a much wider diameter compared to the small intestine. The surface of the large intestine has many small pouches of peritoneum, filled with fat, these are called **omental appendices**. Running longitudinally along the surface of the large intestine are three strips of muscle collectively called the **teniae coli**. They are the **mesocolic, free** and **omental coli**. They contract to shorten the wall of the bowel which produces sacculations known as **haustra**. At the rectosigmoid junction, the smooth muscle of the teniae coli broaden to form a complete layer within the rectum.

Vasculature in the colon

The ascending colon receives arterial supply from two branches of the superior mesenteric artery; the **ileocolic** and **right colic** arteries. The ileocolic artery gives rise to **colic, anterior cecal** and **posterior cecal** branches that supply blood to the ascending colon.

The transverse colon is supplied by the **right colic**, and **middle colic** artery which are branches of the **superior mesenteric** artery and the **left colic** artery which is a branch of the **inferior mesenteric** artery.

A single branch of the **inferior mesenteric** artery called the **left colic** artery supplies blood to the descending colon. The sigmoid colon receives blood supply through the **sigmoid** arteries which are branches of the **inferior mesenteric** artery.

The venous drainage of the colon is similar to its arterial supply. The ascending colon has the **ileocolic** and **right colic** veins, which empty into the **superior mesenteric** vein.

The transverse colon has the **middle colic** vein, which empties into the **superior mesenteric** vein.

The descending colon has the **left colic** vein, which drains into the **inferior mesenteric** vein.

The sigmoid colon has the **sigmoid** veins, which empties into the **inferior mesenteric** vein.

The **superior mesenteric** and **inferior mesenteric** veins empty into the **hepatic portal** vein. This is done to allow toxins absorbed from the colon to be detoxified by the liver.

Rectum

The **rectum** connects the sigmoid colon and terminates into the anal canal. It is a site for temporary storage of feces.

Vasculature of the rectum

The rectum receives blood supply through **superior, middle** and **inferior rectal** arteries. The venous drainage is through the corresponding **superior, middle** and **inferior rectal** veins.

Anal Canal

The **anal canal** is the terminal section of the alimentary canal and plays a role in **defecation**. The anal canal is collapsed by the **internal** and **external anal sphincters** to prevent the passage of fecal material, except during defecation. The **anorectal ring** is a muscular ring at the junction of the rectum and the anal canal formed by the fusion of the internal anal sphincter, external anal sphincter and puborectalis muscle.

Vasculature of the anal canal

The anal canal receives blood supply through the **superior** and **inferior rectal** arteries. The venous drainage is through the corresponding **superior**, and **inferior rectal** veins.

The accessory organs and glands of the digestive system

Salivary glands

There are three pairs of salivary glands: the parotid, sublingual and submandibular glands. The **parotid gland** is a bilateral salivary gland located in the face. This is the largest salivary gland. It produces serous saliva, a watery solution rich in enzymes which lubricates and aids in the breakdown of food. The parotid gland is supplied blood by the **posterior auricular** and **superficial temporal arteries**. While venous drainage is through the retromandibular vein.

The **sublingual glands** are the smallest of the three salivary glands and the most deeply situated. They secrete only about 5% of the total saliva volume. The **sublingual** and **submental** arteries and veins supply and drain the sublingual glands.

The **submandibular glands** are located in the face. Their secretions are a mix of serous and mucous salivary secretions. This is important for the lubrication of food during chewing or **mastication**. The submandibular gland receives blood through the **submental** and **sublingual** artery, while it is drained by the **facial** and **sublingual veins**.

Liver

The **liver** performs many functions including detoxifying the blood, producing bile, metabolism of carbohydrates, fats and proteins, storing iron, blood and vitamins, recycling red blood cells and producing plasma proteins. The liver is covered by a fibrous layer, known as **Glisson's capsule**. It is divided into a **right** and **left** lobe. The right lobe is further divided into the **caudate** and **quadrate lobe** that are separated by a deep, transverse fissure known as the **porta hepatis**.

Bile is the fluid secreted from the liver, and stored in the gallbladder, and released into the duodenum. Bile contains bile salts that are formed from cholesterol. It works to break down fat by emulsification and eliminates products from the breakdown of red blood cells. Bile is secreted from hepatocytes and drains from both lobes of the liver via canaliculi, intralobular ducts and collecting ducts into the left and right hepatic ducts.

The liver has a unique dual blood supply. It receives blood from the **hepatic artery** and the **hepatic vein**. The hepatic artery supplies arterial blood to the liver. The hepatic portal vein is the dominant blood supply to the liver parenchyma and it supplies the liver with partially deoxygenated blood, carrying nutrients absorbed from the small intestine. This allows the liver to perform its gut-related functions, such as detoxification. Blood is drained from the liver through the hepatic veins. The **central veins** of the hepatic lobule form **collecting veins** which then combine to form **multiple hepatic veins**.

Gallbladder

The **gallbladder** functions to concentrate and store bile which is produced by the liver. The stored bile is released from the gallbladder in response to **cholecystokinin**. The gallbladder is typically divided into the **fundus, body** and **neck**. The **biliary tree** is a series of gastrointestinal ducts that allow newly synthesized bile from the liver to be concentrated and stored in the gallbladder. The gallbladder receives blood through the **cystic** artery, and drained by the **cystic vein**.

Pancreas

The **pancreas** is an abdominal glandular organ with both digestive and hormonal functions. We will discuss hormonal functions of the pancreas in chapter 6. The part of the pancreas that plays a role in digestion is lobulated and is a serous gland comprising clusters of cells called acini, connected by short intercalated ducts. The pancreas are divided into five parts, the **head, uncinate process, neck, body** and **tail**. The head connects the pancreas to the duodenum. Secretions of the pancreas into the duodenum are controlled by a valve called the **sphincter of Oddi**. The pancreas are supplied blood by the **splenic artery**. The head is also supplied by the **superior** and **inferior pancreaticoduodenal arteries**. The head of the pancreas drain blood into the **hepatic portal vein**. The rest of the pancreas drain via the **splenic vein**.

Questions to cover the digestive system

In the questions that follow there might be more than one correct answer

1. **Which of the following are parts of the small intestine?**
    a. **Cecum**
    b. **Ileum***
    c. **Duodenum***
    d. **Jejunum***
    e. **Appendix**

Answer: The small intestine consists of three parts. The first part, called the duodenum, connects to the stomach. The middle part is the jejunum. The third part, called the ileum, attaches to the colon

1. **Which of the following organs of the digestive system is the largest visceral organ by mass in the human body and consists of four lobes?**
    a. **Stomach**
    b. **Liver***
    c. **Pancreas**
    d. **Large Intestine**
    e. **Gall bladder**

Answer: The liver is the largest visceral organ by mass and anatomically it has four lobes: right, left, caudate, and quadrate.

1. **Which structure in the digestive system is affected by gastric reflux?**
    a. **Stomach**
    b. **Liver**
    c. **Esophagus***

d. **Pancreas**
e. **Gall bladder**

Answer: Gastroesophageal reflux disease, or GERD, is a digestive disorder that affects the ring of muscle between your esophagus and your stomach. This ring is called the lower esophageal sphincter. This results in stomach acids back-flowing into the esophagus.

1. **Which of the following is the last part of the gastrointestinal tract?**

   a. Sigmoid colon
   b. Rectum
   c. Descending colon
   d. Cecum
   e. Anal canal*

Answer: The anal canal is the terminal section of the alimentary canal and plays a role in defecation. On one end it terminates in the internal and external sphincters and on the other end it is connected to the rectum.

1. **Which of the following is the distal most part of the large intestine?**

   a. Sigmoid colon
   b. Rectum*
   c. Descending colon
   d. Cecum
   e. Anal canal

Answer: The rectum connects the sigmoid colon and terminates into the anal canal. It is a site for temporary storage of feces.

1. **Which part of the digestive tract acts as a site for water absorption from the feces, and a site for flatus to be stored before being expelled?**

a. **Sigmoid colon***
b. **Rectum**
c. **Descending colon**
d. **Cecum**
e. **Anal canal**

Answer: Some carbohydrates are not digested or absorbed in the small intestine. When food is not digested it passes into the large intestine. The undigested food is then broken down into small parts by good bacteria in the colon. This process makes gas (flatus) that is stored in the sigmoid colon. This is also a site of water absorption from the feces..

1. **Into which of the following organs are bile and pancreatic juices secreted?**
   a. **Ileum**
   b. **Stomach**
   c. **Jejunum**
   d. **Duodenum***

Answer: After its passage through the stomach, ingested food turned into acidic chyme arrives in the first segment of the small intestine, a U-shaped tube called the duodenum. The duodenum produces hormones and receives bile and pancreatic juices. This facilitates chemical digestion in the duodenum while also ensuring the acidity of chyme coming from the stomach is neutralized.

1. **Which of the following is a small pouch that sits just under the liver and stores bile produced by the liver?**
   a. **Left Kidney**
   b. **Urinary Bladder**
   c. **Gallbladder***
   d. **Spleen**
   e. **Vermiform Appendix**

Answer: The gallbladder concentrates and stores bile produced by the liver. The biliary tree is a series of gastrointestinal ducts that allow newly synthesized bile from the liver to be concentrated and stored in the gallbladder. The stored bile is released from the gallbladder in response to cholecystokinin.

1. **Which salivary gland is the largest?**

a. Sublingual gland
b. Submandibular gland
c. Parotid gland
d. Thyroid gland
e. Tongue

Answer: There are three pairs of salivary glands: the parotid, sublingual and submandibular glands. The parotid gland is the largest salivary gland and the sublingual glands are the smallest salivary glands..

1. **Which is the gustatory organ of the digestive system?**

a. Salivary glands
b. Liver
c. Stomach
d. Pancreas
e. Tongue*

Answer: The process of tasting termed gustation begins within the taste buds of the oral cavity and is ultimately expressed in the brain. The taste buds of the tongue allow us to detect sweet, sour, salty, bitter, and umami basic tastes.

Chapter 8: The Cardiovascular System.

The circulatory or cardiovascular system delivers oxygen and nutrients throughout the body and transports carbon dioxide and metabolic waste away from the cells. This system can be broken down into three major parts –the heart, blood, and blood circulation. This chapter will review the major functions of the cardiovascular system in those three parts.

The Heart

The two basic functions of the heart are

1) to pump deoxygentated blood from the rest of the body back to the lungs; and

2) to pump oxygenated blood away from the lungs to the rest of the body.

These two functions are performed by two sides of the heart. Since the heart and lungs are anatomically close, blood doesn't have to be pumped large distances for the first function of the heart. For the second job, the heart has to pump blood all throughout the body which is quite a distance and therefore involves more effort. The normal adult human heart pumps about 5 liters of blood every minute throughout life.

Anatomically your heart is approximately the size of your closed fist and lies between your lungs, behind the sternum. It is slightly tilted to the left at its tip. The heart is enclosed in a **pericardial sac**, which contains a small amount of serous fluid that acts to reduce surface tension and lubricate, thereby facilitating the free movement of the heart.

There are three layers of the heart wall. The outer layer of the heart wall is the **epicardium**, the middle layer is the **myocardium**, and the inner layer is the **endocardium**.

The **epicardium** secretes pericardial fluid which protects the tissues as they rub together when a heart beats.

The **myocardium** lies underneath the epicardium and is the second layer of the wall of the heart. It makes up the bulk of the heart and is the layer of cardiac muscle tissue responsible for contraction of the heart.

The **endocardium** is the serous membrane of connective tissues which lines the inside of the heart and is continuous with the blood vessels.

The chambers of the heart

The heart has four chambers - an **atria** and **ventricle** on the right and left side of the heart. The atria are the chambers on the top of the heart and the ventricles are the chambers on the bottom of the heart.

The two atria are thin-walled chambers that receive blood from the veins. Each atrium has a small out-pouching called an **auricle** which increases the volume of the atrium. The lining of each atrium is smooth, except for the anterior atrial walls and the lining of the two auricles, which contain projecting muscle bundles called the **musculi pectinati**, which give the auricles their rough appearance. The two atria are separated from each other by an internal **interatrial septum**.

The two **ventricles** are thick-walled chambers that forcefully pump blood out of the heart. The <u>right atrium receives deoxygenated blood</u> from systemic veins; the <u>left atrium receives oxygenated blood</u> from the pulmonary veins.

Valves of the Heart

Directionality of blood flow in the heart is enforced by its valves. The heart has **two types of valves** that keep the blood flowing in the correct direction. The valves between the atria and ventricles are called atrioventricular valves (also called cuspid valves), while those at the bases of the large vessels leaving the ventricles are called semilunar valves. When the ventricles contract, atrioventricular valves close to prevent blood from flowing back into the atria. When the ventricles relax, semilunar valves close to prevent blood from flowing back into the ventricles.

The **four valves** of the heart are:

The **right atrioventricular valve** is a **tricuspid valve** located between the right atrium and the right ventricle.

Pulmonary valve is a **semilunar valve** located between the right ventricle and the pulmonary artery.

Mitral valve is a **bicuspid valve** located between the left atrium and the left ventricle.

Aortic valve is a **semilunar valve** located between the left ventricle and the aorta.

Pathway of Blood through the Heart

The heart works as two simultaneous pumps, one on the right and one on the left. These pumps fulfill the two functions of the heart that we talked about earlier in this chapter.

Blood flows from the right atrium to the right ventricle, and then is pumped to the lungs to receive oxygen. From the lungs, the blood flows to the left atrium, then to the left ventricle. From there it is pumped to the systemic circulation.

Though we describe above the flow of blood through the right side of the heart and then through the left side, it is important to realize that both atria and ventricles contract at the same time.

The Cardiac Cycle

The cardiac cycle comprises all of the physiological events associated with a single heartbeat. The atria and ventricles alternately contract in each cardiac cycle. The pressures in the chambers change greatly over the course of the cardiac cycle.

The cardiac cycle is essentially split into two phases, systole (the contraction phase) and diastole (the relaxation phase). Each of these is then further divided into an atrial and ventricular component. They occur as the heart beats, pumping blood through a system of blood vessels that carry blood to every part of the body. Systole occurs when the heart contracts to pump blood out, and diastole occurs when the heart relaxes after contraction.

The four stages of a cardiac cycle are:

1. **Atrial systole** is the phase when both atria contract and force the blood from the atria into the ventricles. This phase lasts about 0.1 seconds.
2. During the **ventricular systole** both ventricles contract, and blood is forced to the lungs via the pulmonary trunk, and the rest of the body via the aorta. This phase lasts about 0.3 seconds.
3. **Atrial diastole** is the relaxation phase of the atria, during which the atria fill with blood from the large veins (the vena cavae). This lasts for about 0.7 seconds.
4. The **ventricular diastole** is the last phase of the cardiac cycle and it lasts about 0.5 seconds. It begins before atrial systole, allowing the ventricles to fill passively with blood from the atria.

Blood supply to the heart

The right and left **coronary arteries**, branches of the **ascending aorta**, supply blood to the walls of the myocardium. After blood passes through the capillaries in the myocardium, it enters a system of cardiac or coronary veins. Most of the cardiac veins drain into the coronary sinus, which opens into the right atrium.

Conduction system of the heart

The cardiac conduction system is a collection of nodes and specialized conduction cells that initiate and coordinate contraction of the heart muscle. The cardiac conduction system has a few specialized structures that are described below.

The **sinoatrial (SA) node** or the **pacemaker of the heart** is located in the upper wall of the right atrium, at the junction where the superior vena cava enters. These cells can generate electrical impulses that spreads via gap junctions across both atria, resulting in **atrial contraction** or **atrial systole**, causing the blood to move from the atria into the ventricles.

After the electrical impulses spread across the atria, they converge at the **atrioventricular (AV) node**, located within the atrioventricular septum, near the opening of the coronary sinus. The AV node acts to delay the impulses by approximately 120ms, to ensure the atria have enough time to fully eject blood into the ventricles before **ventricular systole**.

The wave of excitation then passes from the atrioventricular node into the **atrioventricular bundle** or **the bundle of His**. This is a continuation of the AV node tissue, and descends down the membranous part of the interventricular septum, before dividing into two main bundles that conduct electrical impulses into the **Purkinje fibers** of the right and left ventricles.

The Purkinje fibers also called the **sub-endocardial plexus of conduction cells**, are a network of specialized cells located in the subendocardial surface of the ventricular walls, and are able to rapidly transmit cardiac action potentials from the atrioventricular bundle to the myocardium of the ventricles. This rapid conduction allows coordinated ventricular contraction (ventricular systole) and blood is moved from the right and left ventricles to the pulmonary artery and aorta respectively.

The **sequence of electrical events** during one full contraction of the heart muscle are as follows:

1. An excitation signal also called an action potential is created by the **sinoatrial (SA) node**.
2. The wave of excitation spreads across the atrial chambers of the heart, causing them to contract.
3. Upon reaching the **atrioventricular (AV) node**, the excitation signal is delayed.
4. It is then conducted into the **bundle of His**, down the interventricular septum.
5. The **bundle of His** and the **Purkinje fibers** spread the wave impulses along the ventricular chambers of the heart, causing them to contract.

Electrocardiogram

The potential differences between the depolarization and repolarization of the myocardium can be detected using electrodes placed on the skin. This process is called the electrocardiography. The graph of voltage versus time of the electrical activity of the heart is called the electrocardiogram or ECG. There are three main components to an ECG:

1. the **P wave**, which represents the depolarization of the right and left atria;
2. the **QRS complex**, which represents the depolarization of the right and left ventricles; and
3. the **T wave**, which represents the repolarization of the ventricles.

Blood

The second major component of the cardiovascular system is the blood. Flowing around the body, blood carries oxygen and nutrients, collects waste, distributes hormones and dissipates heat. An adult has about 5 liters of blood.

It is a connective tissue, and like all connective tissues, it is made up of cellular elements and an extracellular matrix. The cellular elements include:

1. **red blood cells (RBCs)** or **erythrocytes** which carry oxygen; RBCs are a biconcave disk-shaped cell with no nucleus, each red blood cell contains 300 million **hemoglobin** molecules. **Hemoglobin** is composed of heme, an iron-rich pigment, and globin, ribbonlike

protein chains. Oxygen in the lungs latches onto heme to make **oxyhemoglobin**. In this form, oxygen travels through the bloodstream to all parts of the body.

2. various types of **white blood cells (WBCs)** or leukocytes that are part of the defense system, and

3. **platelets** or **thrombocytes** which are tiny fragments of much larger cells, are involved in the process of blood clotting.

The extracellular matrix, called **plasma**, is a straw-colored fluid, which is 90% water. It suspends the cellular elements and enables them to circulate throughout the body within the cardiovascular system. About 7% of the volume of plasma is made of proteins. These include several proteins that are unique to the plasma called plasma proteins and a much smaller number of regulatory proteins, including enzymes and some hormones. The three major groups of plasma proteins are:

Albumin is the most abundant of the plasma proteins. Because of its abundance, it helps to maintain both blood volume and blood pressure. It functions as binding proteins for fatty acids and steroid hormones. We learnt in chapter 1 that lipids are hydrophobic; but, their binding to albumin enables their transport in the aqueous plasma. Albumin normally accounts for approximately 54% of the total plasma protein volume.

Globulins are the second most common plasma proteins. There are **three main subgroups** known as **alpha, beta,** and **gamma globulins**. The **alpha** and **beta globulins** transport iron, lipids, and the fat-soluble vitamins A, D, E, and K to the cells. Like albumin, alpha and beta globulins also contribute to osmotic pressure. The **gamma globulins** are proteins involved in immunity and are better known as an **antibodies** or **immunoglobulins**. Although other plasma proteins are produced by the liver, immunoglobulins are produced by specialized **white blood cells** known as **plasma cells**. Globulins make up approximately 38% of the total plasma protein volume.

Fibrinogen is the least abundant plasma protein. It is essential for blood clotting, a process described later in this chapter. Fibrinogen accounts for about 7% of the total plasma protein volume.

In addition to proteins, plasma contains a wide variety of other substances. These include various electrolytes, such as sodium, potassium, and calcium ions; dissolved gases, such as oxygen, carbon dioxide, and nitrogen; various organic nutrients, such as vitamins, lipids, glucose, and amino acids; and metabolic wastes. All of these nonprotein solutes combined contribute approximately 1% to the total volume of plasma.

Oxygenated vs deoxygenated blood

Oxygenated blood or **arterial blood** refers to the blood that has been exposed to oxygen in the lungs. The blood flows to the lungs to take up atmospheric oxygen by hemoglobin in the red blood cells. It flows to the left chamber of the heart from the lungs through the pulmonary vein. The **partial pressure of oxygen** in the oxygenated blood is about **100 mmHg**. The oxygenated blood is rich in oxygen as well as other nutrients such as glucose, amino acids, and vitamins. It flows from the heart to the metabolizing tissues throughout the body to supply oxygen and nutrients to the cells. Oxygenated blood flows through the systemic arteries in the body.

Deoxygenated blood or **venous blood** refers to the blood that has a lower oxygen saturation when compared to the blood leaving the lungs. The tissues of the body take up oxygen from the oxygenated blood and return carbon dioxide as a metabolic waste. Thus, the blood that flows out from the tissues consists of a low partial pressure of oxygen and high partial pressure of carbon dioxide. This blood flows through the systemic vein to the right atrium of the heart. It flows to the lungs from the right ventricle to the lungs through the pulmonary artery.

Blood circulation

The third major component of the cardiovascular system is **blood circulation** through blood vessels. These are the channels or conduits through which blood is distributed to body tissues. The vessels make up **two closed systems of tubes** that begin and end at the heart.

1. The **pulmonary vessels**, transports blood from the right ventricle to the lungs and back to the left atrium.
2. The **systemic vessels**, carries blood from the left ventricle to the tissues in all parts of the body and then returns the blood to the right atrium.

Based on their structure and function, blood vessels are classified as either **arteries, capillaries,** or **veins.**

Arteries carry blood away from the heart towards organs and tissues. Apart from the

pulmonary arteries, all arteries also called systemic arteries carry oxygenated blood. Their thick walls and muscular and elastic layers can withstand the high pressure that occurs as the heart pumps blood.

Blood is pumped from the ventricles into large elastic arteries that branch repeatedly into smaller and smaller arteries until the branching results in microscopic arteries called **arterioles.** The arterioles play a key role in regulating blood flow into the tissue capillaries. About 10% of the total blood volume is in the systemic arterial system at any given time.

An arterial wall has three layers.

1. The **tunica intima** (also called **tunica interna**), is the innermost layer. It is comprised of simple **squamous epithelium** surrounded by a connective tissue **basement membrane** with elastic fibers.
2. The middle layer, the **tunica media**, is primarily **smooth muscle** and is usually the thickest layer. It not only provides support for the vessel but also changes vessel diameter to regulate blood flow and blood pressure.
3. The outermost layer, is the **tunica externa** or **tunica adventitia.** This attaches the blood vessel to the surrounding tissue. This layer is connective tissue with varying amounts of elastic and collagenous fibers.

Capillaries, are the smallest and most numerous of the blood vessels. The capillaries convey blood between arteries and veins. Capillaries have a vital role in the exchange of gases, nutrients, and metabolic waste products between the blood and the tissue cells. Substances pass through the capillary wall by diffusion, filtration, and osmosis. Oxygen and carbon dioxide move across the capillary wall by diffusion.

A typical capillary is about 0.01 mm in diameter, only slightly wider than a red blood cell. Smooth muscle cells in the arterioles where they branch to form capillaries regulate blood flow from the arterioles into the capillaries.

Many capillaries enter tissue to form a capillary bed, a site where capillaries link small arteries called **arterioles** to small veins called **venules**. This is the site where oxygen and other nutrients are released, and where waste matter passes into the blood. The thin capillary wall allows easy movement of substances between surrounding tissues. **Fluid movement** across a capillary wall is determined by a combination of **hydrostatic** and **osmotic pressure**. The net result of the capillary microcirculation created by hydrostatic and osmotic pressure is that substances leave the blood at one end of the capillary and return at the other end.

Tissues such as skeletal muscle, liver, and kidney have extensive capillary networks because they are metabolically active and require an abundant supply of oxygen and nutrients. Other tissues, such as connective tissue, have a less abundant supply of capillaries. The epidermis of the skin and the lens and cornea of the eye completely lack a capillary network.

Veins carry blood toward the heart. Apart from the pulmonary veins, all veins also called systemic veins carry deoxygenated blood. The walls of veins have the same three layers as the arteries. Although all the layers are present, there is less smooth muscle and connective tissue. This makes the walls of veins thinner than those of arteries, which is related to the fact that blood in the veins has less pressure than in the arteries. Because the walls of the veins are thinner and less rigid than arteries, veins can hold more blood. Almost 70% of the total blood volume is in the veins at any given time. Medium and large veins have venous valves, similar to the semilunar valves associated with the heart, that help keep the blood flowing toward the heart.

Pulse and Blood Pressure

Pulse refers to the rhythmic expansion of an artery that is caused by ejection of blood from the ventricle. **Blood pressure** commonly refers to arterial blood pressure, the pressure in the aorta and its branches. Systolic pressure is due to ventricular contraction and is 120 mmHg. Diastolic pressure occurs during cardiac relaxation and is 80 mmHg. Pulse pressure is the difference between systolic pressure and diastolic pressure and is 40 mmHg.

Four major factors affect blood pressure. They are

1. **Cardiac output**, which is the amount of blood the heart pumps in 1 minute.

2. **Blood volume**, which is determined by the amount of water and sodium ingested, excreted by the kidneys into the urine, and lost through the gastrointestinal tract, lungs and skin.
3. **Peripheral resistance** is the resistance of the arteries to blood flow. As the arteries constrict, the resistance increases and as they dilate, resistance decreases.
4. **Blood viscosity** which is mainly determined by the **haematocrit** which is the ratio of the volume of red blood cells to the total volume of blood; and **plasma viscosity** which is determined by the plasma protein levels.

When any of these four factors increase, blood pressure also increases. **Pressure receptors** called **baroreceptors**, located in the walls of the large arteries in the thorax and neck, are important for short-term blood pressure regulation.

Chapter 9: The Respiratory System.

The respiratory system works with the circulatory system to provide oxygen and to remove the waste products of metabolism. It also helps to regulate pH of the blood. The term respiration can mean three things. On the macroscopic level it could simply mean **ventilation** or **breathing**. **External respiration** refers to the gas exchange between the blood and the air; while **internal respiration** refers to the gas exchange between the blood and the cells. Cellular respiration refers to the biochemical process through which cells utilize oxygen for metabolism and yield carbon dioxide as a waste product.

The organs involved in ventilation can be divided into the **upper** and **lower respiratory tracts**. The upper respiratory tracts includes the following structures.

Nose and nasal cavity. The nose consists of its **outer part** which is visible and the **nasal cavity**. The nasal cavity has a bony skeleton and is lined with **mucus membrane** and the **olfactory epithelium**. The olfactory epithelium has the smell receptors and helps in processing the sense of smell. In the lower part of the nose the respiratory meatus begins, and it provides mechanical cleansing of the inspired air. It consists of the **inferior meatus** or **lower conchae**. This is the space between the floor of the nasal cavity and the **inferior turbinate**. This is the largest of the air spaces. The **nasolacrimal duct** or **tear duct** empties any drainage from the eyes into the inferior meatus.

The head of the **nasal wall, inferior meatus,** and **osseous piriform aperture** make up the **nasal valve**. The **middle meatus** is the nasal passageway that lies below the inferior meatus and is the site of drainage for three of the **paranasal sinuses**: the **maxillary, frontal,** and **anterior ethmoid sinuses**. Airflow through the paranasal sinuses also creates the tones of our voices. The superior meatus or upper conchae is the nasal space that lies below the middle meatus. Two of the **paranasal sinuses**: the **sphenoid** and **posterior ethmoid sinuses** drain into this passageway.

The **pharynx**, commonly called the throat, is a passageway that extends from the base of the skull to the level of the sixth cervical vertebra. It serves both the respiratory and digestive systems by receiving air from the nasal cavity and air, food, and water from the oral cavity. The pharynx is divided into three regions according to location:

1. the **nasopharynx**, a portion of the pharynx that is posterior to the nasal cavity and extends inferiorly to the uvula.
2. the **oropharynx**, a portion of the pharynx that is posterior to the oral cavity; and
3. the **laryngopharynx** or **hypopharynx**, a portion of the pharynx that extends from the hyoid bone down to the lower margin of the larynx

The **larynx** is a flexible passageway for air between the **oropharynx** and **trachea**, which plays an essential role in voice formation and protects the lower airway against food inspiration.

The larynx is often divided into three sections: **sublarynx**, **larynx**, and **supralarynx**. It is formed by nine cartilages that are connected to each other by muscles and ligaments.

The **trachea** or **windpipe** is a short, flexible air tube, which extends from the larynx down to the middle of the thorax, where it divides into two main bronchi. It allows the passage of air from the larynx and pharynx to the lungs. The **hyaline cartilage** in the tracheal wall provides support and keeps the trachea from collapsing. The mucus membrane that lines the trachea is ciliated pseudostratified columnar epithelium similar to that in the nasal cavity and nasopharynx. **Goblet cells** produce mucus that traps airborne particles and microorganisms, and the cilia propel the mucus upward, where it is either swallowed or expelled.

The lower respiratory tracts includes the following structures.

The **bronchi** are airway passages that conduct air from the trachea to the lungs. The bronchi branch and form the **bronchial tree**. The first two bronchi arising from the trachea are the right and left main bronchi. Each main bronchus divides into **lobar bronchi**, the lobar bronchus divides into **segmental bronchi**, further, into **intrasegmental bronchi**. Finally, the smallest bronchi divide into **bronchioles**. Each bronchiole divides until the **terminal bronchiole**, which continues as the **pulmonary alveolus**, forming the **alveolar tree** within the lungs. The alveolar ducts and alveoli consist primarily of simple squamous epithelium, which permits rapid diffusion of oxygen and carbon dioxide. Exchange of gases between the air in the lungs and the blood in the capillaries occurs across the walls of the alveolar ducts and alveoli.

The **lungs** are the most prominent organs of the respiratory system. Lungs are the final organ of the lower respiratory tract providing the inner and outer respiration and gas exchange between blood and inhaled air. There are two lungs in the human body, the right and the left lung, conical-shaped organs situated in the thoracic cavity. The right lung is divided into three lobes, while the left lung consists of two lobes. Each lobe further divides into bronchopulmonary segments, which is an anatomical and functional unit of the lung.

The lungs are soft and spongy because they are mostly air spaces surrounded by the alveolar cells and elastic connective tissue. They are separated from each other by the **mediastinum**, which contains the heart. The only point of attachment for each lung is at the **hilum**, or **root**, on the medial side. This is where the bronchi, blood vessels, lymphatics, and nerves enter the lungs.

The **right lung is shorter**, **broader**, and has a **greater volume** than the **left lung**. It is divided into three lobes and each lobe is supplied by one of the secondary bronchi. The left lung is longer and narrower than the right lung. It has an indentation, called the **cardiac notch**, on its medial surface for the apex of the heart. The left lung has two lobes.

Each lung is enclosed by a double-layered serous membrane, called the **pleura**. The visceral pleura is firmly attached to the surface of the lung. At the hilum, the visceral pleura is continuous with the parietal pleura that lines the wall of the thorax. The small space between the visceral and parietal pleurae is the **pleural cavity**. It contains a thin film of serous fluid that is produced by the pleura. The fluid acts as a lubricant to reduce friction as the two layers slide against each other, and it helps to hold the two layers together as the lungs inflate and deflate.

Mechanics of Ventilation

Inspiration is the phase of ventilation in which air enters the lungs. **Passive inspiration** is initiated by **contraction** of the inspiratory muscles the **diaphragm** and the **external intercostal muscles**. **Active** or **forced inspiration** involves accessory muscles in addition to the muscles involved in passive inspiration. The accessory muscles are **scalenes, sternocleidomastoid,**

pectoralis major, **pectoralis minor**, **serratus anterior** and the **latissimus dorsi**.

The action of the inspiratory muscles results in an increase in the volume of the thoracic cavity. As the lungs are held against the inner thoracic wall by the pleural seal, they also undergo an increase in volume. An increase in lung volume results in a decrease in the pressure within the lungs. The pressure of the environment external to the lungs is now greater than the environment within the lungs, therefore air moves into the lungs down the pressure gradient.

Expiration is the phase of ventilation in which air is expelled from the lungs. It is initiated by **relaxation** of the inspiratory muscles the **diaphragm** and the **external intercostal muscles**. **Active** or **forced expiration** involves accessory muscles in addition to the muscles involved in passive expiration. The accessory muscles are the **anterolateral abdominal wall, internal intercostal** and the **innermost intercostal muscles**.

The action of the inspiratory muscles results in a decrease in the volume of the thoracic cavity. As the lungs are held against the inner thoracic wall by the pleural seal, they also undergo a decrease in volume. A decrease in lung volume results in an increase in the pressure within the lungs. The pressure of the environment external to the lungs is now lower than the environment within the lungs, therefore air moves out the lungs down the pressure gradient.

Respiratory Volumes and Capacities

An instrument called a **spirometer** is used to measure the volume of air that moves into and out of the lungs, and the process of taking the measurements is called **spirometry**. There are four respiratory volumes:

Tidal volume is the volume of air that enters and leaves with each breath, from a normal quiet inspiration to a normal quiet expiration.

Inspiratory reserve volume is the extra volume of air that can be inspired above tidal volume, from normal quiet inspiration to maximum inspiration.

Expiratory reserve volume is the extra volume of air that can be expired below tidal volume, from normal quiet expiration to maximum expiration.

Residual or **reserve volume** is the volume of air remaining after maximum expiration.

Capacities are composites of two or more lung volumes. They are fixed as they do not change with the pattern of breathing.

Vital capacity or **forced vital capacity** is the volume that can be exhaled after maximum inspiration to maximum expiration. It is the sum of inspiratory reserve volume, tidal volume and expiratory reserve volume.

Inspiratory capacity is the volume breathed in from quiet expiration to maximum inspiration. It is the sum of inspiratory reserve volume and tidal volume.

Functional residual capacity is the volume remaining after quiet expiration. It is the sum of expiratory reserve volume and residual volume.

Total lung capacity is the volume of air in lungs after maximum inspiration. It is the sum of all volumes.

Anatomical dead space is the volume of air that never reaches alveoli and so never participates in respiration. It includes volume in upper and lower respiratory tract up to and including the terminal bronchioles.

Chapter 10: Urinary System.

The urinary system is part of the excretory system of the body which also includes the respiratory, integumentary, and digestive systems. The purpose of the urinary system is to eliminate wastes from the body, regulate blood volume and pressure, control levels of electrolytes and metabolites, and regulate blood pH.

Components of the Urinary System

The urinary system consists of the **kidneys, ureters, urinary bladder**, and **urethra**. The kidneys form the urine and account for the other functions attributed to the urinary system. The ureters carry the urine away from kidneys to the urinary bladder, which is a temporary reservoir for the urine. The urethra is a tubular structure that carries the urine from the urinary bladder to the outside. The female and male urinary system are very similar, differing only in the length of the urethra.

Kidneys

The kidneys are bilateral bean-shaped, reddish-brown organs located in shallow depressions against the posterior abdominal wall and behind the parietal peritoneum. Their main function is to filter and excrete waste products from the blood. They are also responsible for water and electrolyte balance in the body. The other components of the urinary system are accessory structures to eliminate the urine from the body. The right kidney is usually slightly lower than the left because the liver displaces it downward. Each kidney is encased in complex layers of fascia and fat. They are held in place by connective tissue, called **renal fascia** or **Gerota's fascia**, and is surrounded by a thick layer of adipose tissue, called **perirenal fat**, which helps to protect it. A tough, fibrous, connective tissue **renal capsule** closely envelopes each kidney and provides support for the soft tissue that is inside.

Internal Organization of the Kidney

The renal parenchyma can be divided into two main areas – the **outer cortex** and **inner medulla**. The cortex extends into the medulla, dividing it into triangular shapes called **renal pyramids**.

The apex of a renal pyramid is called a **renal papilla**. Each renal papilla is associated with a structure known as the **minor calyx**, which collects urine from the pyramids. Several minor calices merge to form a **major calyx**. Urine passes through the major calices into the renal pelvis, and from there the urine drains into the ureter, which transports it to the bladder for storage.

The medial margin of each kidney is marked by a deep fissure, known as the **renal hilum**. The renal vessels and ureter enter/exit the kidney via the renal hilum.

Nephron

The nephron is the basic functional unit of a kidney. It consists of three parts:

1. the **renal corpuscle**, the filtering component. This consists of a cluster of capillaries, called the **glomerulus**, surrounded by a double-layered epithelial cup, called the **glomerular capsule** or **Bowman's capsule**. An afferent arteriole leads into the renal corpuscle and an efferent arteriole leaves the renal corpuscle.
2. the **renal tubule**, which is responsible for absorption and ion secretion. The renal tubule can be divided into three components known as the **proximal convoluted tubule**, the **Loop of Henle** and the **distal convoluted tubule**.
3. the **collecting duct**, which is responsible for the final reabsorption of water and for storing urine.

The kidneys contain two types of nephrons, superficial cortical nephrons (70-80%) and juxtamedullary nephrons (20-30%). These names refer to the location of the glomerular capsule, which is either in the outer cortex of the kidney, or near the corticomedullary border. Urine passes from the nephrons into collecting ducts then into the minor calyces. The **juxtaglomerular apparatus**, which monitors blood pressure and secretes renin, is formed from modified cells in the afferent arteriole and the ascending limb of the nephron loop.

Ultrafiltration in the Glomeruli

Blood filters into the Bowman's capsule in a process called ultrafiltration, a filtration that occurs under pressure. The afferent arteriole dilates, while the efferent arteriole constricts. This creates a pressure gradient throughout the glomerulus, causing filtration under pressure. The filtration rate of molecules of the same charge across the filtration barrier is inversely related to their molecular weight. Small molecules like glucose are freely filtered whereas large molecules like albumin are barely able to cross the barrier. Also, negatively charged large molecules filter less easily than positively charged ones of the same size.

Counter-current multiplication in the Kidneys

Counter-current multiplication is the process that allows the kidneys to produce concentrated urine. It occurs in the loop of Henle, a hairpin-like structure comprised of a thin descending limb, a thin ascending limb and a thick ascending limb within the nephron. As the thick ascending limb is impermeable to water, the interstitium becomes concentrated with ions, increasing its osmolarity. This drives water reabsorption from the descending limb as water moves from areas of low osmolarity to areas of high osmolarity. This system is known as counter-current multiplication and it allows the kidneys to reabsorb around 99% of filtered water. Majority of the counter-current multiplication occurs in the loops of Henle of juxtamedullary nephrons.

Vasculature of the kidneys

The kidneys are supplied blood through the **renal arteries** which arise from the abdominal aorta. The renal artery enters the kidney via the renal hilum. At the hilum level, the renal artery forms an **anterior** and a **posterior division**, which carry 75% and 25% of the blood supply to the kidney, respectively. **Five segmental arteries** originate from these two divisions. The **avascular plane of the kidney (line of Brodel)** is an imaginary line which delineates the segments of the kidney supplied by the anterior and posterior divisions. The segmental arteries undergo further divisions to supply the renal parenchyma. Each segmental artery divides to form **interlobar arteries** that are situated either side of every renal pyramid. The interlobar arteries undergo further division to form the **arcuate arteries** which divides further to form the **interlobular arteries.**

The interlobular arteries pass through the cortex, dividing one last time to form **afferent arterioles.** The afferent arterioles form a capillary network, the **glomerulus**, where filtration takes place. The capillaries come together to form the **efferent arterioles**. In the outer two-thirds of the renal cortex, the efferent arterioles form what is a known as a **peritubular network**, supplying the nephron tubules with oxygen and nutrients. The inner third of the cortex and the medulla are supplied by long, straight arteries called **vasa recta**. The kidneys are drained of venous blood by the **left** and **right renal veins**. They leave the renal hilum anteriorly to the renal arteries, and empty directly into the **inferior vena cava.**

Ureters

Ureters are two thick tubes which transport urine from the kidney to the bladder. Each ureter is connected to one kidney. These muscular tubes are about 28 cm in length, beginning at the renal pelvis. They transport urine from the kidney to the bladder.

The wall of the ureter consists of three layers.

1. The **outer layer** or the **fibrous coat**, is a supporting layer of fibrous connective tissue.
2. The **middle layer** or the **muscular coat**, consists of the **inner circular** and **outer longitudinal smooth muscle** that are involved in peristalsis to propel the urine.
3. The **inner layer** or the **mucosa**, is lined with **transitional epithelium** that **secretes mucus**, which coats and protects the surface of the cells.

The arterial supply to the ureters is from the **renal artery, testicular/ovarian artery**, and **ureteral branches** of the **abdominal aorta** and the **superior** and **inferior vesical arteries**. Venous drainage is from the veins corresponding to the arteries mentioned above.

Urinary Bladder

The bladder is a hollow organ of the urinary system located in the pelvic cavity. It is the site of temporary storage reservoir of urine and is involved in voiding. The inner lining of the urinary bladder is a mucous membrane of transitional epithelium that is continuous with that in the ureters. When the bladder is empty, the mucosa has numerous folds called **rugae**. The rugae and transitional epithelium allow the bladder to expand as it fills. It can accommodate about 600ml of urine in healthy adults.

There are three openings in the floor of the urinary bladder. Two of the openings are from the **ureters**. Small flaps of mucosa cover these openings and act as valves that allow urine to enter the bladder but prevent it from backing up from the bladder into the ureters. The is the opening into the **urethra**. A band of the **detrusor muscle** encircles this opening to form the **internal urethral sphincter**.

The vasculature of the bladder is primarily derived from the **internal iliac vessels**. Arterial supply is via the **superior vesical branch** of the **internal iliac artery**. In males, this is supplemented by the **inferior vesical artery**, and in females by the **vaginal arteries**. In both sexes, the **obturator** and **inferior gluteal arteries** also contribute small branches. Venous drainage is achieved by the **vesical venous plexus**, which empties into the **internal iliac veins**. The vesical plexus in males is in continuity with the **prostate venous plexus** (plexus of Santorini), which also receives blood from the **dorsal vein** of the penis.

Urethra

The **urethra** is the tube that transports urine from the floor of the urinary bladder to an **external urethral orifice**. It is a thin-walled tube lined by **stratified columnar epithelium**, which is protected from the corrosive urine by mucus secreting glands. The wall also contains smooth muscle fibers and is supported by connective tissue.

Two sphincters control the flow of urine through the urethra. They are:.

1. The **internal urethral sphincter** made of smooth (involuntary) muscle at the beginning of the urethra, where it leaves the urinary bladder.
2. The **external urethral sphincter** made of skeletal (voluntary) muscle encircles the urethra where it goes through the pelvic floor.

In females, the urethra is short, about 4 cm long. The external urethral orifice opens to the outside just anterior to the opening for the vagina.

In males, the urethra is longer, about 20 cm in length, and transports both urine and semen. The first part, next to the urinary bladder, passes through the prostate gland and is called the **prostatic urethra**. The second part, is called the **membranous urethra** and it penetrates the pelvic floor and enters the penis. The third part, the **spongy urethra**, is the longest region. This portion of the urethra extends the entire length of the penis, and the external urethral orifice opens to the outside at the tip of the penis.

Functions of the Urinary System

Micturition is the process of eliminating water and electrolytes from the urinary system, commonly known as urinating. It consists of three phases that require coordinated relaxation of the bladder and urethral sphincters, which are controlled by the sympathetic, parasympathetic and somatic nervous systems. The three phases of micturition are:

1. **Storage/continence phase**, when urine is stored in the bladder. Storage of urine requires relaxation of the **detrusor muscle** of the bladder, and simultaneous contraction of both the **internal urethral sphincters** and **external urethral sphincters**. The **bladder** and **internal urethral sphincters** are under the control of the sympathetic nervous system while the **external urethral sphincters** is under the control of the somatic nervous system.

2. **Voiding phase**, where urine is released through the urethra. This phase is initiated voluntarily. This process requires contraction of the **detrusor muscle** of the bladder, and simultaneous relaxation of both the **internal urethral sphincters** and **external urethral sphincters**. This is coordinated via the **spinal-pontine-spinal reflex** which involves the **micturition control center** in the **pons**.

3. **Termination of voiding phase**, where micturition ends. As voiding of the bladder is completed the urine flow reduces and ends. The **external urethral sphincters** close under voluntary control. The **urethra contracts** forcing urine above the level of the external sphincter back up into the bladder. The **micturition control center** in the **pons** takes control again and allows the filling cycle to start again.

Regulate blood pH

The urinary system alters blood pH by

1. Excretion of hydrogen (H^+) ions as
   a. dihydrogen phosphate, or
   b. ammonia; and

2. Production of bicarbonate (HCO_3^-) ions; and

3. Reabsorption of bicarbonate (HCO_3^-) ions

1. **Excretion of Hydrogen (H^+) Ions**

    a. **Excretion of H^+ ions in the form of dihydrogen phosphate ($H_2PO_4^-$). H^+ ions are actively transported into the lumen via hydrogen-ATPase pumps on alpha intercalated cells present in the late distal convoluted tubule, the connecting segment, cortical and outer medullary collecting duct, and the early part of the inner medullary collecting duct. Excess luminal phosphate binds a large portion of hydrogen ions, buffering them as $H_2PO_4^-$ before excretion. This excretion of H^+ ions increases blood pH.**

    b. **Excretion of H^+ ions in the form of ammonium (NH_4^+). Glutamine is converted to glutamate and ammonium in the proximal convoluted tubule (PCT). The ammonium dissociates to ammonia and H^+ ions, allowing it to pass the membrane and enter the lumen. Once in the lumen, it reforms ammonium by picking up a luminal H^+ ion. This allows hydrogen to be excreted as ammonium (NH_4^+) ions, increasing blood pH.**

1. **Production of bicarbonate (HCO_3^-) ions.** The metabolic activity of cells in the kidney produces large amounts of carbon dioxide. This then reacts with water to produce **bicarbonate (HCO_3^-) ions**, which enter the plasma, and **H^+ ions** that is transported into the lumen via the **sodium/hydrogen (Na^+/H^+) exchanger**. These H^+ ions drive HCO_3^- reabsorption in the **proximal convoluted tubule** (PCT). Bicarbonate can also be produced from amino acids, which produces ammonium ions which then enter the urine.

1. **Reabsorption of bicarbonate (HCO_3^-) ions.** H^+ ions are transported into the lumen via the **sodium/hydrogen (Na^+/H^+) exchanger** and combine with any **bicarbonate.** On the luminal side **carbonic anhydrase** drives the formation of **carbonic acid (H_2CO_3).** Carbonic acid then dissociates into **carbon dioxide (CO_2)** and **water (H_2O),** which both can diffuse into the cell. Once inside the cell, **carbonic anhydrase** converts **carbon dioxide (CO_2)** and **water (H_2O)** into **carbonic acid (H_2CO_3).** This dissociates into **bicarbonate (HCO_3^-) ions** which is transported into the blood, and **H^+ ions** that are transported back into the lumen for the cycle to repeat.

Chapter 11: Lymphatic System.

The lymphatic system functions to drain tissue fluid, plasma proteins and other cellular debris back into the blood stream. This fluid is called **interstitial fluid** and is found in spaces around cells. This interstitial fluid helps bring oxygen and nutrients to cells and to remove waste products from them. As new interstitial fluid is made, it replaces older fluid, which drains towards lymph vessels. When it enters the lymph vessels, it is called lymph. Both lymph and interstitial fluid resemble blood plasma in composition. The other functions of the lymphatic system are the **absorption of fats and fat-soluble vitamins** from the digestive system and the subsequent transport of these substances to the venous circulation. The third and probably most well-known function of the lymphatic system is **defense against invading microorganisms and disease.** Lymph nodes and other lymphatic organs filter the lymph to remove microorganisms and other foreign particles. Lymphatic organs contain lymphocytes that destroy invading organisms.

Components of the Lymphatic System

The lymphatic system consists of

1. **Lymph**
2. **Lymphatic vessels that transport the lymph, and**
3. **Lymphoid organs/tissues**

Lymph

Lymph is a fluid similar to blood plasma and is made up of water, glucose, protein, fats, and salt. It is derived from blood plasma as fluids pass through capillary walls at the arterial end. As the interstitial fluid begins to accumulate, it is picked up and removed by tiny lymphatic vessels and returned to the blood. As soon as the interstitial fluid enters the lymph capillaries, it is called lymph. Returning the fluid to the blood prevents edema and helps to maintain normal blood volume and pressure.

Lymphatic Vessels

Lymphatic vessels carry lymph but unlike blood vessels they only carry fluid away from the tissues. The smallest lymphatic vessels are the **lymph capillaries,** which begin in the tissue spaces as blind-ended sacs. The wall of the lymph capillary is composed of endothelium in which the simple squamous cells overlap to form a simple one-way valve. This arrangement permits fluid to enter the capillary but prevents lymph from leaving the vessel.

Small lymphatic vessels join to form larger tributaries, called **lymphatic trunks**, which drain large regions. Lymphatic trunks merge until the lymph enters the **two lymphatic ducts**.

1. **Right lymphatic duct** drains lymph from the upper right quadrant of the body.
2. **Thoracic duct** drains all the rest.

There is **no pump** in the lymphatic system like the heart. The pressure gradients to move lymph through the vessels come from the skeletal muscle action, respiratory movement, and contraction of smooth muscle in vessel walls.

Lymphoid Organs/Tissues

Lymphoid organs have clusters of lymphocytes and other cells, such as macrophages, enmeshed in a framework of short, branching connective tissue fibers. They can be divided into **primary** or **secondary lymphoid organs**.

Primary or **central lymphoid organs** generate lymphocytes from progenitor cells. These include the

1. **Bone marrow** is a semi-solid tissue found within the spongy or cancellous portions of the bone. It is the *site of T and B cell generation* and *B cell maturation*.
2. **Thymus** is a soft organ with two lobes that is located anterior to the ascending aorta and posterior to the sternum. It is relatively large in infants and children but after puberty it begins to decrease in size. This is the **site for T cell maturation**. The thymus also produces a hormone, **thymosin**, which stimulates the maturation of lymphocytes in other lymphoid organs.

Secondary or **peripheral lymphoid organs** maintain mature naïve lymphocytes and are the sites for **lymphocyte activation by antigens**. These include the

1. **Lymph Nodes** filter substances that travel through the lymphatic fluid, and they contain lymphocytes (white blood cells) that help the body fight infection and disease. They are enclosed in a fibrous capsule and is made up of an **outer cortex, paracortex** and **inner medulla** and divided into compartments called **nodules.** The outer cortex consists of groups of inactivated B cells called follicles. The deeper paracortex consists of the T cells interacting with dendritic cells. The medulla contains large blood vessels, sinuses and medullary cords that contain antibody-secreting plasma cells, macrophages, and B cells.

2. **Tonsils** are clusters of lymphoid tissue under the mucus membrane that line the nose, mouth and throat. They are also called **Waldeyer's tonsillar ring.** Lymphocytes and macrophages in the tonsils provide protection against harmful substances and pathogens that may enter the body through the nose or mouth. There are 4 groups of tonsils in the **Waldeyer's tonsillar ring** and they are

    a. **One Pharyngeal tonsil located near the opening of the nasal cavity into the pharynx.**

    b. **Two Tubal tonsils are located near the opening of the auditory tubes into the nasopharynx.**

    c. **Two Palatine tonsils are located near the opening of the oral cavity into the pharynx.**

    d. **Several Lingual tonsils are located on the posterior surface of the tongue, near the opening of the oral cavity into the pharynx.**

3. **Spleen** is an intraperitoneal organ located in the upper left quadrant of the abdomen, under cover of the diaphragm and the ribcage. In fetuses, the spleen produces blood cells. In adults, the spleen has other roles and is made of **red pulp** and **white pulp**, separated by the **marginal zone.**

    a. **Red pulp is the site of mechanical filtration of red blood**

cells to remove particulate matter and aged red blood cells. The red pulps has several different types of blood cells, including platelets, granulocytes, red blood cells, and plasma.

    b. **White pulp is the site of active immune response and can be divided into two regions**

        i. The **periarteriolar lymphoid sheaths** (PALS) contain T lymphocytes and are typically around the arteriole supply of the spleen.

        ii. Between the PALS and the border of the red pulp are lymph follicles with dividing B lymphocytes. They are a site of antibody production.

    c. **Marginal zone is the area between the red and white pulp. It is located farther away from the central arteriole. It contains dendritic cells and macrophages.**

Fat absorption function of the lymphatics

Lacteals are **lymphatic capillaries** in the villi of the small intestine. **Chylomicrons** or **ultra-low density lipoproteins** contain triglycerides, phospholipids, cholesterol and proteins. They are formed in the small intestine and taken up by the lacteals. This enable fats and cholesterol to move within the bloodstream.

Immune function of the lymphatics

We had briefly touched upon the immune functions of the lymphatic system while we described the role of various lymphoid organs. White blood cells, known as lymphocytes are formed in the bone marrow. There are two types of lymphocyte, T cells and B cells. An easy way to remember this is to consider that B cells mature in the bone marrow and T cells mature in the thymus. This is why both bone marrow and thymus are considered **primary lymphoid organs.**

T and B cell types

T Helper Cells assist other lymphocytes, including maturation of B cells.

Cytotoxic T Cells destroy virus-infected cells and tumor cells, and are involved in transplant rejection

Memory T Cells' clones expressing a specific T cell receptor can persist for decades in our body. This allows the body to respond to the same antigen when encountered at a later time.

Regulatory T Cells are immunosuppressive and shut down T-cell mediated-immunity towards the end of an immune response.

Plasma Cells are long-lived, non-proliferating antibody-secreting B cells.

Memory B Cells' clones expressing a specific B cell receptor can persist for decades in our body. This allows the body to respond to the same antigen when encountered at a later time.

Regulatory B Cells are immunosuppressive and stops the expansion of pathogenic, pro-inflammatory lymphocytes. They also activate regulatory T cells

Circulating lymphocytes pass through lymph nodes. These are immune-surveillance sites of the body. If the lymph fluid derived from blood plasma has viruses, bacteria, or other foreign particles they act as antigens and activate the lymphocytes in lymph nodes. The activated lymphocytes can engage in **cell-mediated immune response** and **antibody-mediated immune response**.

Cell-mediated immune response incorporates the

1. Activation of macrophages and natural killer cells enabling them to destroy intracellular pathogens;
2. T-cell mediated immunity through cytotoxic T cells that helps destroy cells that show cell surface markers of infection
3. Cytokine secretion by activated cells to modulate immune signaling.

Antibody-mediated or **humoral immune response** incorporates these steps

1. Activation of B cells to process and present antigens on the cell surface by MHC Class II proteins.
2. Binding of T helper cells to these surface expressed, MHC Class II protein-associated antigens.
3. Activation of T helper cells after antigen binding and secretion of cytokines that induce B cell proliferation.

4. B cells proliferation and secretion of antibodies against the antigen that was initially expressed on the cell surface in Step 1.

Chapter 12: Nervous System.

The nervous system is one of two organ systems in the body that maintains homeostasis. The other is the endocrine system. The response from the nervous system is fast but usually short acting while the response from the endocrine systems tends to be slower but long acting. We have covered homeostasis in Chapter 2 and endocrine system in Chapter 6. In this chapter we will consider the structure and function of the nervous system. This can be broadly organized as the **central nervous system** and the **peripheral nervous system.**

Functional divisions of the nervous system.

The nervous system has two major anatomical divisions.

1. The **central nervous system (CNS)** consists of the brain and the spinal cord. It integrates and coordinates the processing of sensory data and the transmission of motor commands. The CNS is also the seat of higher functions, such as intelligence, memory, and emotion.
2. The **peripheral nervous system (PNS)** connects the CNS with the rest of the body. The PNS consists of two divisions.
    a. **Afferent division** brings sensory information into the CNS through sensory **receptors** in the body. We covered sensory receptors in the integumentary system in Chapter 5.
    b. **Efferent division** carries motor commands from the CNS to the rest of the body. The target organs and tissues that respond to the motor commands from the CNS are called **effectors**. The effector division can be subdivided into two systems.
        i. The **somatic nervous system (SNS)** controls the skeletal muscle contractions which in turn can be **voluntary** or **reflexive**. Voluntary contractions are under conscious control while reflexive contractions are involuntary or automatic.
        ii. The **autonomic nervous system (ANS)** or **visceral motor system** controls the smooth and cardiac muscles, glandular secretions and tissues without a

conscious effort. This system has three subdivisions

1. **Sympathetic division** controls the "fight-or-flight" responses and therefore prepares the body for stress like by increasing the heart rate.

2. **Parasympathetic division** controls the "rest-and-digest" responses and therefore prepares the body to recover from stress like by lowering the heart rate.

3. **Enteric nervous system** regulates the movement of water and electrolytes between the gut lumen and tissue fluid compartments by directing the activity of secretomotor neurons that innervate the mucosa in the small and large intestines. In doing so it modulates gastric, immune and endocrine functions.

Cells of the Nervous System

There are two main types of cells in the nerve tissue.

1. **Neuron** is the actual nerve cell. It is conductive and transmits impulses. It is the structural unit of the nervous system. They are highly specialized and amitotic. Therefore, if a neuron is destroyed, it cannot be replaced because neurons do not go through mitosis.

2. **Neuroglia** or **glial cells** are the support cells for nerves. They are far more numerous than neurons. They are nonconductive and do not transmit nerve impulses. Glial cells nourish, and protect the neurons, and are capable of mitosis.

Structure of a neuron

Each neuron has three basic parts: **cell body (soma)**, **dendrites**, and **axon.**

Cell Body

The cell body of a neuron is similar to other types of cells in many aspects. It has a nucleus, nucleolus and other cytoplasmic organelles normally found in other cells. The one striking difference is that mature neurons lack centrioles. Centrioles are involved in cell division, but since neurons are amitotic they do not undergo cell division and therefore have no need for centrioles.

Dendrites

Dendrites are branched cytoplasmic extensions that project from the cell body. They can also be referred as fibers. They are usually, short, branching and numerous, which increases their surface area to receive signals from other neurons. They are called **afferent processes** because they transmit impulses to the neuron cell body.

Axon

Axons also are branched cytoplasmic extensions that project from the cell body. They are usually elongated. An axon may have infrequent branches called **axon collaterals**. Axons and axon collaterals terminate in many short branches or **telodendria**. The distal ends of the telodendria are slightly enlarged to form **synaptic bulbs**. Axons are called **efferent processes** because they transmit impulses away from the neuron cell body.

Some axons are surrounded by a segmented, white, fatty substance called the **myelin sheath**. The unmyelinated regions between the myelin segments are called **the nodes of Ranvier**. Myelinated neurons usually transmit impulses faster than unmyelinated neurons. The white matter in the CNS is made up of myelinated fibers, while unmyelinated fibers and cell bodies make the gray matter.

Differences in myelin synthesis between PNS and CNS

Schwann cells produce myelin in the PNS. The cytoplasm, nucleus, and outer cell membrane of the Schwann cell forms a tight covering around the myelin and around the axon itself at the nodes of Ranvier. This covering is called the **neurilemma**, and unlike the axon and myelin sheath the neurilemma does not degenerate after a a nerve has been cut or crushed. The hollow tube of the neurilemma aids the nerve fibers in the PNS to regenerate.

Oligodendrocytes or **oligodendroglia** are specialized glial cells that wrap themselves around neurons present in the CNS. Oligodendrocytes are primarily responsible for maintenance and generation of the myelin sheath that surrounds axons.

Neurons can be classified based on both its structure and function.

Structural classification of neurons

Based on its anatomy neurons can usually classified as one of the following:

1. **Unipolar neurons** usually have only a single cytoplasmic extension called a neurite which can then branch to form dendritic and axonal processes. Unipolar neurons are typically seen in invertebrates.
2. **Bipolar neurons** have only two cytoplasmic extensions, one of which is the axon and the other a dendrite. They are normally found in the afferent division of the PNS as part of the sensory pathways.
3. **Multipolar neurons** have many cytoplasmic extensions, one of which is the axon and the rest are dendrites. Most neurons in the CNS are multipolar neurons.
4. **Anaxonic neurons** have many cytoplasmic extensions but one cannot distinguish the dendrites from the axon. Such neurons are found in the retina.
5. **Pseudounipolar neurons** have one axon that projects from the cell body for relatively a very short distance, before splitting into two branches one that extends to the CNS and the other to the PNS. They develop embryologically as a bipolar neuron.

Functional classification of neurons

Based on its function neurons can usually classified into one of the following three classes based on the direction in which they transmit impulses relative to the CNS.

1. **Afferent**, or **sensory neurons** carry impulses from peripheral sense receptors to the CNS. They usually have long dendrites and relatively short axons.
2. **Efferent**, or **motor neurons** transmit impulses from the CNS to effector organs such as muscles and glands. Efferent neurons usually have short dendrites and long axons.
3. **Interneurons**, or **association neurons**, are located entirely within the CNS in which they form the connecting link between the afferent and efferent neurons. They have short dendrites and may have either a short or long axon.

Types of neuroglia

There are six types of neuroglia four of which are found in the CNS and two in the PNS.

1. **Neuroglia in the CNS**
    a. **Astrocytes are the most abundant glial cells in the CNS. They have numerous projections which help in clinging to the neurons and capillaries forming the blood-brain barrier. They regulate the chemical environment around the neurons.**
    b. **Microglial cells, are specialized macrophages in the CNS and are involved in both immune response and maintaining homeostasis. The phagocytose debris, dead cells and foreign materials.**
    c. **Ependymal cells, or ependymocytes secrete cerebrospinal fluid. The ciliary action of these cells helps in CSF circulation in the CNS.**
    d. **Oligodendrocytes coat neurons in the CNS with myelin sheath that insulates the axon and allows electrical signals to propagate more efficiently.**

1. **Neuroglia in the PNS**
    a. **Schwann cells are similar to oligodendrocytes in function. They coat neurons in the PNS with myelin sheath that insulates the axon and allows electrical signals to propagate more efficiently.**
    b. **Satellite cells are small cells that surround the cell body of neurons in the PNS. The regulate the microenvironment of the neurons.**

The Central Nervous System

The brain and spinal cord in the dorsal body cavity comprise the CNS. The brain is surrounded by the cranium, and the spinal cord is protected by the vertebrae. The brain is continuous with the spinal cord at the **foramen**

magnum. The CNS is surrounded by connective tissue membranes, called **meninges**, and by **cerebrospinal fluid.**

Meninges

There are **three layers of meninges** around the brain and spinal cord.

1. The **dura mater** is the outer layer made of tough white fibrous connective tissue.
2. The **arachnoid** is the middle layer of meninges and resembles a cobweb in appearance. It is a thin layer with numerous threadlike strands that attach it to the innermost layer. The space under the arachnoid, the **subarachnoid space**, is filled with **cerebrospinal fluid** and contains **blood vessels.**
3. The **pia mater** is the innermost layer of meninges. This thin, delicate membrane composed of fibrous tissue is tightly bound to the surface of the brain and spinal cord. It is pierced by blood vessels that travel to the brain and spinal cord.

Brain

The brain is divided into the **cerebrum, diencephalons, brain stem**, and **cerebellum.**

Cerebrum

Cerebrum is the largest and most obvious portion of the brain. It is divided by a deep longitudinal fissure into two cerebral hemispheres. The two hemispheres are two separate entities but are connected by a band of white fibers, called the **corpus callosum** that provides a communication pathway between the two halves.

Each cerebral hemisphere is divided into five lobes, four of which have the same name as the bone over them: the **frontal lobe**, the **parietal lobe**, the **occipital lobe**, and the **temporal lobe**. A fifth lobe, the **insula** or **Island of Reil**, lies deep within the **lateral sulcus**. The lateral sulcus is a deep fissure in each hemisphere that separates the frontal and parietal lobes from the temporal lobe.

Diencephalons

The **diencephalons** are centrally located and is nearly surrounded by the cerebral hemispheres. It includes the

1. **Thalamus**, which is the largest component of the diencephalons. It consists of two oval masses of gray matter that serve as relay stations for sensory impulses, except for the sense of smell, going to the cerebral cortex.
2. **Hypothalamus**, is located below the thalamus. It plays a key role in maintaining homeostasis because it regulates many visceral activities.
3. **Epithalamus** is a small gland involved with the onset of puberty and rhythmic cycles in the body. It is like a biological clock.

Brain Stem

The brain stem is the region between the diencephalons and the spinal cord. It consists of three parts:

1. **Midbrain**, is the most superior portion of the brain stem. It's functions is to integrate motor, sensory, and cognitive performances between the cerebral cortex on one side of the brain to the same region on the other side.
2. **Pons**, is the bulging midportion of the brain stem. It contains nuclei that relay signals from the forebrain to the cerebellum, along with nuclei that deal primarily with sleep, respiration, swallowing, bladder control, hearing, equilibrium, taste, eye movement, facial expressions, facial sensation, and posture.
3. **Medulla oblongata**, or **medulla**, extends inferiorly from the pons. It is continuous with the spinal cord at the **foramen magnum**. All the ascending (sensory) and descending (motor) nerve fibers connecting the brain and spinal cord pass through the medulla.

Cerebellum

The **cerebellum**, is the second largest portion of the brain and, is located below the occipital lobes of the cerebrum. Three paired bundles of myelinated nerve fibers, called **cerebellar peduncles**, form communication pathways between the cerebellum and other parts of the central nervous system.

Ventricles and Cerebrospinal Fluid

The **ventricular system** is a set of communicating cavities within the brain. These structures are responsible for the production, transport and removal of **cerebrospinal fluid**, which bathes the central nervous system..

Spinal Cord

The **spinal cord** extends from the foramen magnum at the base of the skull to the level of the first lumbar vertebra. Like the brain, the spinal cord is surrounded by bone, meninges, and cerebrospinal fluid.

The spinal cord is divided into **31 segments: 8 cervical, 12 thoracic, 5 lumbar, 5 sacral** and **1 coccygeal**. A pair of spinal nerves leaves each segment of the spinal cord. The spinal cord has two main functions:

1. **Conduction pathway** for impulses going to and from the brain. Sensory impulses travel to the brain on ascending tracts in the cord. Motor impulses travel on descending tracts.
2. **Reflex center.** The reflex arc is the functional unit of the nervous system. Reflexes are responses to stimuli that do not require conscious thought and consequently, they occur more quickly than reactions that require thought processes.

Action Potential

Action potentials are nerve signals. Neurons generate and conduct these signals to other neurons or target tissues. This is how information is passed in the nervous system. Physiologically, an action potential is caused by temporary changes in membrane permeability for diffusible ions. These changes cause ion channels to open and the ions to decrease their concentration gradients.

An action potential has several phases.

1. **Hypopolarization** is the initial increase of the membrane potential to the value of the threshold potential.
2. **Depolarization** is the next phase when the threshold potential opens voltage-gated sodium channels and causes a large influx of sodium ions.
3. The inside of the neuron now becomes more and more electropositive, until the potential gets closer the electrochemical equilibrium for sodium of +61 mV. This phase of extreme positivity is

the **overshoot phase**.

4. In the **repolarization phase**, the sodium channels close leading to the sudden decrease in sodium permeability. The overshoot value of the cell potential opens voltage-gated potassium channels, which causes a large potassium efflux, decreasing the cell's electropositivity. The purpose of this phase is to restore the resting membrane potential.

5. Repolarization always leads first to **hyperpolarization**, a state in which the membrane potential is more negative than the default membrane potential. But soon after that, the membrane establishes again the values of membrane potential.

6. The **refractory period** is the time after an action potential is generated, during which the excitable neuron cannot produce another action potential.

Chapter 13: Review of Anatomy

In chapters 2 -12 we covered one anatomical system per chapter. In this chapter we will highlight the key concepts of each chapter.

Ch 2. Homeostasis

1. **Homeostasis** is a condition where the body resists change in order to maintain a stable, relatively constant internal environment.
2. **Five parts** of a **homeostatic system** are:
    a. **Stimulus,**
    b. **Detector,**
    c. **Regulator,**
    d. **Effector** and
    e. **Response.**
3. **Stimulus** is a physical, chemical or environmental factor that causes deviation from the normal body environment.
4. A **detector** or **sensor** receives the stimulus and forwards it to the control center.
5. The **regulator** or **control center** receives and processes information from the sensor. It also sets the normal reference point for any physiological processes.
6. The **effector** is the cell, tissue or organ that responds to the signals from the regulator.
7. **Response** is the corrective measure taken by the effector toward the stimulus.
8. There are **two types of feedback mechanism** in a homeostatic system.
    a. **Negative feedbacks** counteract the change of a body property from its target value and usually dampen the effect of the initial stimuli.
    b. **Positive feedback** on usually amplifies their initiating stimuli.

Case Study. Our sweating response

When you exercise, your muscles increase heat production, nudging your body temperature upward. Let's consider the steps involved in bringing the temperature back to normal of around 98.6°F (37°C)

1. Increased metabolic activity of the muscles during exercises leads to an increase in temperature.
2. Hypothalamus is the body's thermostat.
3. The hypothalamus has temperature receptor cells which detect changes in the temperature of the blood flowing through the brain.
4. Neurons from the hypothalamus send electrical signals to sweat glands in the skin,
5. Sweat glands secrete sweat.
6. Evaporation of the sweat cools you off bringing the body temperature back to normal.

In the above case can you identify the five components **stimulus, detector, regulator, effector** and **response** of the homeostatic system?

a. **Stimulus**; Increase in body temperature
b. **Detector**; Thermoreceptors in the hypothalamus
c. **Regulator**; Hypothalamus
d. **Effector**; Sweat glands
e. **Response**; Sweating

Ch 3. The Skeletal System.

1. **Six functions** of the **skeletal system** are:
   a. Support the body;
   b. Protect soft organs;
   c. Generate movement;
   d. Blood formation or hematopoiesis;
   e. Fat storage; and
   f. Mineral storage.
2. **Two types** of **skeletal system** are:
   a. **Axial skeleton** forms the central axis of the body and

consists of bones which form the **skull, vertebral column,** and **thoracic cage.**

   b. **Appendicular skeleton** forms the appendages of the body and connects it to the axial skeletal system. It consists of bones which form the **pectoral** and **pelvic girdles,** the **limb bones,** and the bones of the **hands** and **feet.**

3. **Six types of bones** in the **skeletal system** are:

a. **Flat bones** provide protection like a shield and because these bones are somewhat flattened provide large areas of attachment for muscles.

b. **Long bones** are longer than they are wide and support the weight of the body and allow movement along with the muscular system.

c. **Short bones** are about as long as they are wide and provide stability and some movement.

d. **Irregular bones** vary in shape and structure and often have a fairly complex shape, which helps protect internal organs.

e. **Sutural bones** are small, irregular bones within the cranial sutures. They are also called **Wormian** bones.

f. **Sesamoid bones** are small, round bones embedded in tendons and they reinforce and decrease stress on that tendon.

1. **Different types of joints**

   a. Based on structure there are 3 types of joints;

      i. **Synovial joints** are characterized by the presence of a fluid-filled joint cavity contained within a fibrous capsule.

      ii. **Fibrous joints** have thick connective tissues composed of collagen fibers found between the articulations of the joints.

      iii. **Cartilaginous joints** are skeletal joints that are entirely joined by cartilage.

   b. Based on function there are 3 types of joints;

      i. **Synarthrosis** is a type of join that is strong and immovable like the skull sutures.

      ii. **Amphiarthrosis** joints allow slight movement like

the pubic symphysis of the pelvic girdle.

    iii. **Diarthrosis** is a type of joint that allows full movement like the elbow.

2. **There are 4 types of cells in the bone**

    a. **Osteoblasts** are single-nucleated cells that are responsible for the synthesis and mineralization of bone during both initial bone formation (called ossification) and later bone remodeling.

    b. **Osteocyte** is an osteoblast surrounded by a calcified matrix that it secreted.

    c. **Osteoclasts** are specialized multinucleated giant cells that resorb bone.

    d. **Bone lining cells** cover inactive or non-remodeling bone surfaces.

Ch 4. The Muscular System.

1. **Three functions** of the **muscular system** are:

    a. Generate motion;

    b. Support body posture; and

    c. Heat production.

2. **Three types of muscles** are:

    a. **Skeletal muscle** is striated and is under voluntary control of the somatic nervous system.

    b. **Cardiac muscles** are striated, found only in the heart wall and are controlled involuntarily by the autonomic nervous system.

    c. **Smooth muscles** are non-striated and are controlled involuntarily by the autonomic nervous system.

3. Muscle contractions are defined by changes in the length of the muscle during contraction. There are **five types of muscle contractions**

    a. In an **isometric contraction** the length of the muscle does not change.

    b. In an **isotonic contraction** the length of the muscle

changes.

   c. An isotonic contraction where the muscle shortens is called a **concentric contraction.**

   d. An isotonic contraction where the muscle lengthens is called an **eccentric contraction.**

   e. A muscle contraction in which the velocity of the muscle contraction remains constant while the length of the muscle changes is called an **isokinetic contraction.**

4. Muscles contract through the process described by the **sliding filament model.** It is a cycle of repetitive events that causes actin and myosin myofilaments to slide over each other, contracting the sarcomere and generating tension in the muscle.

5. **Types of Natural Lever Systems**

   a. **First class lever**, the fulcrum is in the middle of the effort and the load.

   b. **Second class lever**, the load is in the middle between the fulcrum and the effort.

   c. **Third class lever**, the effort is in the middle between the fulcrum and the load.

Ch 5. The Integumentary System.

The integumentary system is the largest organ of the body that forms a physical barrier between the external environment and the internal environment that it serves to protect and maintain.

1. Seven functions of the integumentary system are :

   a. Barrier function;

   b. Detection of stimuli;

   c. Body temperature regulation;

   d. Allow movement;

   e. Cell fluid maintenance;

   f. Protect against microorganisms; and

   g. Synthesis of Vitamin D

2. Three layers of skin are

   a. **Epidermis;** the topmost layer of the skin

b. **Dermis,** is the middle layer of the integument.

c. **Hypodermis,** is below the dermis.

3. Cells of the epidermis are:

   a. Largely composed of **keratinocytes** undergoing terminal maturation.

   b. **Melanocytes** that produce melanin and pigment formation.

   c. **Langerhans cells** that are antigen-presenting cells of the immune system.

   d. **Merkel cells** that are the sensory mechanoreceptors.

4. **Five layers** of the epidermis are:

   a. **Stratum basale** is the innermost layer and it has keratinocytes undergoing mitosis in this layer.

   b. **Stratum spinosum** contains cells called prickle cells that have small radiating processes that connect with other cells. Keratin is synthesized in this layer.

   c. **Stratum granulosum** contains cells that secrete lipids and other waterproofing molecules.

   d. **Stratum lucidum** cells have lost their nuclei and have increased keratin production.

   e. **Stratum corneum** is the outermost layer of cells that have lost all organelles and have been hardened with keratin.

5. **Two layers** of the dermis are:

   a. The **superficial papillary layer,** is composed of loose connective tissue that is highly vascular; and

   b. The deeper **reticular layer** has blood vessels, connective tissue and accessory structures.

6. Accessory structures of the skin in the dermis are:

   a. **Fibroblasts**

   b. **Mast cells**

   c. **Blood vessels,**

   d. **Cutaneous sensory nerves,**

   e. **Hair follicles,**

   f. **Nails,**

   g. **Sebaceous glands;** and

   h. **Sweat glands.**

Ch 6. The Endocrine System.

1. **Hormones** are chemicals secreted by a gland, and they usually affect distant target tissue.
2. There are different types of hormones based on their chemical structure: **lipid-derived, fatty acid-derived, amino acid-derived,** and **peptide** and **protein** hormones.
3. Lipid-derived hormones can diffuse across the plasma membrane.
4. Amino acid- and peptide-derived hormones can not diffuse across the plasma membrane.
5. **Group I hormones** bind to intracellular receptors and are lipid derived.
6. **Group II hormones** bind to cell surface receptors and are hydrophilic.
7. **Major endocrine glands** and **their secretions** are as follows:
   a. **Pituitary gland**
      i. The **adenohypophysis** portion of the pituitary secretes the following **six** hormones
         1. **HGH** or **Human Growth Hormone**
         2. **TSH** or **Thyroid Stimulating Hormone.**
         3. **ACTH** or **Adenocorticotropic Hormone**
         4. **PRL** or **Prolactin**
         5. **FSH** or **Follicle Stimulating Hormone**
         6. **LH** or **Luteinizing Hormone**
      ii. The **neurohypophysis** portion of the pituitary secretes the following **two** hormones.

1. **ADH** or **Antidiuretic Hormone**
2. **Oxytocin**
   a. Thyroid secretes the following **three** hormones.
      i. **T3** or **Triodothyronine.**
      ii. **T4** or **Tetraiodothyronine**
      iii. **Calcitonin**
   b. **Parathyroid** secretes the **PTH** or **Parathyroid Hormone**

c. **Adrenal gland**
   i. The **adrenal cortex** portion of the adrenal gland secretes **three classes** of hormones.
      1. **Mineralocorticoids**
      2. **Glucocorticoids**
      3. **Adrenal androgens**
   ii. The **adrenal medulla** portion of the adrenal gland secretes **amine hormones** like **epinephrine** and **norepinephrine.**

1. The **mixed glands** of the endocrine system are glands with both endocrine and exocrine function
2. Endocrine secretion of mixed glands are as follows
   a. **Islets of Langerhans** of the **Pancreas** secretes the following **three** hormones.
      i. **Glucagon**
      ii. **Insulin**
      iii. **Somatostatin**
   b. **Thymus** secretes **thymosin.**
   c. **Pineal gland** secretes **melatonin**

Ch 7. The Digestive System.

1. The **two basic processes** of digestion are:
   a. Chemical and mechanical breakdown of the food; and
   b. Absorption of the constituents of the food.
2. **Stomach** is the major site of chemical digestion of proteins and some fats, but very little absorption of the digested food happens here.
3. The stomach has four main anatomical divisions
   a. **Cardia,**
   b. **Fundus,**
   c. **Body** and
   d. **Pylorus.**
4. **Small intestine** is a site chemical digestion and absorption of the digested food

5. The small intestine has three main anatomical divisions
    a. **Duodenum,**
    b. **Jejunum,** and
    c. **Ileum**
6. **Colon,** or the **large intestine** is the major site for extracting water and electrolytes from the undigested remains of the food and form feces.
7. The large intestine has four main anatomical divisions
    a. **Ascending colon,**
    b. **Transverse colon,**
    c. **Descending colon** and
    d. **Sigmoid colon.** At its distal end the sigmoid colon connects to the **rectum.**
8. **Rectum** is a site for temporary storage of feces.
9. **Anal canal** is the terminal section of the alimentary canal and plays a role in **defecation.**
10. The **accessory organs** and **glands** of the digestive system include
    a. **Salivary glands**: the parotid, sublingual and submandibular glands.
    b. **Liver;** secretes bile
    c. **Gallbladder;** stores bile and
    d. **Pancreas;** secretes enzymes that break down sugars, fats, and starches.

Ch 8. The Cardiovascular System.

1. The heart has **two** basic functions:
    a. Pump **deoxygentated blood** from the rest of the body back **to the lungs;** and
    b. Pump **oxygenated blood** away from the lungs **to the rest of the body.**
2. The heart wall has **three** layers:
    a. The outer layer of the heart wall is the **epicardium,**
    b. The middle layer is the **myocardium,** and
    c. The inner layer is the **endocardium.**

3. The heart has **four** chambers:
    a. Top chambers: Right and left **atria** receive blood from the veins
        i. The right atrium receives deoxygenated blood from systemic veins
        ii. The left atrium receives oxygenated blood from the pulmonary veins.
    b. Bottom chambers: Right and left **ventricle** pump blood out of the heart
4. The heart has **four valves**:
    a. **Right atrioventricular valve** is a **tricuspid valve** located between the right atrium and the right ventricle.
    b. **Pulmonary valve** is a **semilunar valve** located between the right ventricle and the pulmonary artery.
    c. **Mitral valve** is a **bicuspid valve** located between the left atrium and the left ventricle.
    d. **Aortic valve** is a **semilunar valve** located between the left ventricle and the aorta.
5. **Directionality of blood flow in the heart:**
    a. Blood flows from systemic circulation into the right atrium
    b. From the right atrium blood is pumped to the right ventricle
    c. Then it is pumped to the lungs to receive oxygen.
    d. From the lungs, the blood flows to the left atrium,
    e. Then it is pumped to the left ventricle.
    f. From there it is pumped to the systemic circulation.
6. The **four stages** of a **cardiac cycle** are:
    a. **Atrial systole;** both atria contract and force the blood from the atria into the ventricles.
    b. **Ventricular systole;** both ventricles contract, and blood is forced to the lungs via the pulmonary trunk, and the rest of the body via the aorta.
    c. **Atrial diastole;** relaxation phase of the atria, during which the atria fill with blood from the vena cavae.
    d. **Ventricular diastole;** last phase of the cardiac cycle during which the ventricles fill passively with blood from the atria.

7. **Conduction pathway of the heart** is as follows:
    a. **Sinoatrial (SA) node** creates an action potential.
    b. The wave of excitation spreads across the atrial chambers of the heart, causing them to contract.
    c. The excitation signal is delayed at the **atrioventricular (AV) node**
    d. Action potential is then conducted into the **bundle of His.**
    e. The **bundle of His** and the **Purkinje fibers** spread the electrical signal along the ventricular chambers of the heart, causing them to contract.
8. Three components of an **electrocardiogram** or **ECG** are:
    a. **P wave,** shows the depolarization of the right and left atria;
    b. **QRS complex,** shows the depolarization of the right and left ventricles; and
    c. **T wave,** shows the repolarization of the ventricles.
9. **Blood** has cellular and extracellular elements.
10. **Cellular elements** of blood are:
    a. Red blood cells
    b. White blood cells
    c. Platelets
11. **Extracellular elements** of blood are called plasma
12. **Plasma** contains **proteins** and **non-protein solutes**
13. **Proteins** of plasma include
    a. Albumin
    b. Alpha globulin
    c. Beta globulin
    d. Gamma globulin
    e. Fibrinogen
14. **Non-protein solutes** of plasma include
    a. Electrolytes, such as sodium, potassium, and calcium ions;
    b. Dissolved gases, such as oxygen, carbon dioxide, and nitrogen;
    c. Organic nutrients, such as vitamins, lipids, glucose, and amino acids; and
    d. Metabolic wastes.

15. **Two types** of **blood circulation systems** are
    a. **Pulmonary vessels**, transports blood from the right ventricle to the lungs and back to the left atrium.
    b. **Systemic vessels**, carries blood from the left ventricle to the tissues in all parts of the body and then returns the blood to the right atrium.
16. There are **three types of blood vessels**
    a. **Arteries** carry oxygenated blood away from the heart towards organs and tissues. <u>Important exception:</u> Pulmonary arteries carry deoxygenated blood from the right side of the heart to the lungs.
    b. **Veins** carry deoxygenated blood towards the heart from the organs and tissues. <u>Important exception:</u> Pulmonary veins carry oxygenated blood from the lungs to the left side of the heart.
17. **Capillaries**, are the smallest and most numerous of the blood vessels. They convey blood between arteries and veins. The thin capillary wall allows easy movement of substances between surrounding tissues.
18. **Four** major **factors affect blood pressure**. They are
    a. **Cardiac output**, which is the amount of blood the heart pumps in 1 minute.
    b. **Blood volume**, which is determined by the amount of water and sodium.
    c. **Peripheral resistance**, which is the resistance of the arteries to blood flow
    d. **Blood viscosity** which is mainly determined by the ratio of the volume of red blood cells to the total volume of blood.

Ch 9. The Respiratory System.

1. **Respiration** can mean **three** things.
    a. It could simply mean **ventilation** or **breathing**.
    b. **External respiration** refers to the gas exchange between the blood and the air
    c. **internal respiration** refers to the gas exchange between the

blood and the cells.

2. **Cellular respiration** refers to the biochemical process through which cells utilize oxygen for metabolism and yield carbon dioxide as a waste product.
3. **Pharynx** is the tube connecting the **nose/mouth** to the **esophagus.**
4. **Larynx** is the tube forming a passage between the **pharynx** and **trachea.**
5. **Trachea** is the tube connecting the **larynx** to the **bronchi** of the **lungs.**
6. **Bronchi** are airway passages that conduct air from the **trachea** to the **lungs.**
7. **Alveoli** are sites of **gas-exchange** in the **lungs.**
8. **Diaphragm** is a thoracic muscle that lays beneath the lungs and aids in ventilation.
9. **Spirometer** is used to measure the volume of air that moves into and out of the lungs, and the process of taking the measurements is called **spirometry.**
10. There are four respiratory volumes:
    a. **Tidal volume** is the volume of air that enters and leaves with each breath, from a normal quiet inspiration to a normal quiet expiration.
    b. **Inspiratory reserve volume** is the extra volume of air that can be inspired above tidal volume, from normal quiet inspiration to maximum inspiration.
    c. **Expiratory reserve volume** is the extra volume of air that can be expired below tidal volume, from normal quiet expiration to maximum expiration.
    d. **Residual** or **reserve volume** is the volume of air remaining after maximum expiration.
11. **Vital capacity** or **forced vital capacity** is the volume that can be exhaled after maximum inspiration to maximum expiration.
12. **Inspiratory capacity** is the volume breathed in from quiet expiration to maximum inspiration.
13. **Functional residual capacity** is the volume remaining after quiet expiration.

14. **Total lung capacity** is the volume of air in lungs after maximum inspiration.

15. **Anatomical dead space** is the volume of air that never reaches alveoli and so never participates in respiration.

Ch 10. The Urinary System.

10.1. **Four** functions of the urinary system are:

a. Eliminate wastes from the body;

b. Regulate blood volume and pressure;

c. Control levels of electrolytes and metabolites; and

d. Regulate blood pH.

10.2. The urinary system organs include the

a. Kidneys,

b. Ureter,

c. Bladder, and

d. Urethra.

10.3. Nephrons are the main functional component of the kidneys.

10.4. Three parts of the nephron are:

a. **Renal corpuscle**, the filtering component.

◈ **Glomerulus**, is the basic filtering unit of the kidney consisting of a cluster of capillaries.

b. **Renal tubule**, which is responsible for absorption and ion secretion.

c. **Collecting duct**, which is responsible for the final reabsorption of water and for storing urine.

10.5. **Ultrafiltration** is the filtration under pressure that occurs in the kidneys.

10.6. **Counter-current multiplication** is the process that allows the kidneys to produce concentrated urine.

10.7. **Micturition** is the process of eliminating water and electrolytes from the urinary system and has the following **three** phases.

a. **Storage/continence phase**, when urine is stored in the bladder.

b. **Voiding phase**, where urine is released through the urethra.

c. **Termination of voiding phase**, where micturition ends.

10.8. The urinary system **alters blood pH** by **three** processes

a. Excretion of hydrogen (H^+) ions as

◈ dihydrogen phosphate, or

◈ ammonia; and

b. Production of bicarbonate (HCO_3^-) ions; and

c. Reabsorption of bicarbonate (HCO_3^-) ions

Ch 11. The Lymphatic System.

1. **Three functions** of the **lymphatic system** are:
    a. Drain tissue fluid, plasma proteins and other cellular debris back into the blood stream.
    b. Absorption of fats and fat-soluble vitamins from the digestive system

c.   Defense against invading microorganisms and disease.

2. **Lymph** is a fluid similar to blood plasma and is made up of water, glucose, protein, fats, and salt. It is derived from blood plasma as fluids pass through capillary walls at the arterial end.

3. **Lymphatic vessels** carry lymph but unlike blood vessels they only carry fluid away from the tissues. Like veins lymphatic vessels have valves.

4. There is **no pump** in the lymphatic system like the heart. The pressure gradients to move lymph through the vessels come from the skeletal muscle action, respiratory movement, and contraction of smooth muscle in vessel walls.

5. **Primary** or **central lymphoid organs** generate lymphocytes from progenitor cells. These include the

   a.   **Bone marrow** is the *site of T and B cell generation* and *B cell maturation*.

   b.   **Thymus** is is the **site for T cell maturation** and production of thymosin

6. **Secondary** or **peripheral lymphoid organs** maintain mature naïve lymphocytes and are the sites for **lymphocyte activation by antigens**. These include the

   a.   **Lymph Nodes** that filter substances that travel through the lymphatic fluid.

7. **Tonsils**, which are clusters of lymphoid tissue under the mucus membrane that line the nose, mouth and throat.

8. **Spleen**, which is the site of mechanical filtration of red blood cells to remove particulate matter and aged red blood cells; and the site of active immune response.

9. Types of **T and B cells**

   a.   **T Helper Cells** assist other lymphocytes, including maturation of B cells.

   b.   **Cytotoxic T Cells** destroy virus-infected cells and tumor cells, and are involved in transplant rejection

   c.   **Memory T Cells'** clones expressing a specific T cell receptor can persist for decades in our body. This allows the body to respond to the same antigen when encountered at a later time.

   d. **Regulatory T Cells** are immunosuppressive and shut down T-cell mediated-immunity towards the end of an immune response.

   e. **Plasma Cells** are long-lived, non-proliferating antibody-secreting B cells.

   f. **Memory B Cells'** clones expressing a specific B cell receptor can persist for decades in our body. This allows the body to respond to the same antigen when encountered at a later time.

   g. **Regulatory B Cells** are immunosuppressive and stops the expansion of pathogenic, pro-inflammatory lymphocytes. They also activate regulatory T cells

10. Key steps of **Cell-mediated immune response** are

   a. Activation of macrophages and natural killer cells;

   b. T-cell mediated immunity through cytotoxic T cells

   c. Cytokine secretion by activated cells.

11. Key steps of **Antibody-mediated** or **humoral immune response** are

   a. Activation of B cells to process and present antigens on the cell surface.

   b. Binding of T helper cells to these surface expressed antigens.

   c. Activation of T helper cells after antigen binding and secretion of cytokines that induce B cell proliferation.

   d. B cells proliferation and secretion of antibodies against the antigen.

Ch 12. The Nervous System.

1. The nervous system has two major anatomical divisions.

   a. The **central nervous system (CNS)** consists of the brain and the spinal cord. It integrates and coordinates the processing of sensory data and the transmission of motor commands.

   b. The **peripheral nervous system (PNS)** connects the CNS with the rest of the body.

2. **Two divisions** of the PNS are:.
    a. **Afferent division** brings sensory information into the CNS through sensory **receptors** in the body.
    b. **Efferent division** carries motor commands from the CNS to the rest of the body.
3. **Two sub-divisions** of the **effector division** are

a. The **somatic nervous system (SNS)** controls the skeletal muscle contractions which in turn can be **voluntary** or **reflexive**. Voluntary contractions are under conscious control while reflexive contractions are involuntary or automatic.
b. The **autonomic nervous system (ANS)** or **visceral motor system** controls the smooth and cardiac muscles, glandular secretions and tissues without a conscious effort.

1. **Three subdivisions** of the **ANS** are

a. **Sympathetic division** controls the "fight-or-flight" responses.
b. **Parasympathetic division** controls the "rest-and-digest" responses.
c. **Enteric nervous system** regulates the movement of water and electrolytes between the gut lumen and tissue fluid compartments..

1. **Neurons** are the functional cells of the nervous system. They conduct and transmit electrical impulses.
2. **Neurons** have a **cell body or soma** and cytoplasmic extensions called dendrite and axons.
3. **Dendrites** transmit impulses to the neuron cell body.
4. **Axons** transmit impulses away from the neuron cell body.
5. **Neuroglia** or **glial cells** are the support cells for nerves. They are nonconductive and do not transmit electrical iimpulses.
6. **Schwann cells** produce myelin sheath around nerves in the PNS.
7. **Oligodendrocytes** produce myelin sheath around nerves in the CNS.
8. **Meninges** cover the brain and spinal cord
9. The brain is divided into the

a. **Cerebrum**,

b. **Diencephalons**,

c. **Brain stem**, and

d. **Cerebellum**.

10. The spinal cord is divided into **31 segments** from which a pair of spinal nerves leave:

    a. **8 cervical**,

    b. **12 thoracic**,

    c. **5 lumbar**,

    d. **5 sacral** and

    e. **1 coccygeal**.

11. The **spinal cord** has **two** main **functions**:

    a. **Conduction pathway** for impulses going to and from the brain.

    b. **Reflex center** of the nervous system.

12. **Action potentials** are nerve signals caused by temporary changes in membrane permeability for diffusible ions.

13. An action potential has several phases.

    a. **Hypopolarization;**

    b. **Depolarization;**

    c. **Overshoot phase;.**

    d. **Repolarization phase;**

    e. **Hyperpolarization;** and

    f. **Refractory period**